YAWNS FREEZE YOUR BRAIN

Also by Mick O'Hare

Does Anything Eat Wasps?

Why Don't Penguins' Feet Freeze?

How To Fossilise Your Hamster

Do Polar Bears Get Lonely?

How to Make a Tornado

Why Can't Elephants Jump?

Why Are Orangutans Orange?

Farts Aren't Invisible

YAWNS FREEZE YOUR BRAIN

**More Mind-Blowing Facts
From Science, History,
Life and the Universe**

MICK O'HARE

Bedford Square
Publishers

First published in the UK in 2024 by Bedford Square Publishers Ltd,
London, UK

bedfordsquarepublishers.co.uk
@bedsqpublishers

© Mick O'Hare, 2024

The right of Mick O'Hare to be identified as the author of
this work has been asserted in accordance with the Copyright,
Designs and Patents Act 1988. All rights reserved. No part of this
book may be reproduced, stored in or introduced into a retrieval
system, or transmitted, in any form or by any means (electronic,
mechanical, photocopying, recording or otherwise) without the
written permission of the publishers.

Any person who does any unauthorised act in relation to this
publication may be liable to criminal prosecution and civil
claims for damages.

A CIP catalogue record for this book is
available from the British Library.

ISBN
978-1-83501-140-9 (Paperback)
978-1-83501-141-6 (eBook)

2 4 6 8 10 9 7 5 3 1

Typeset in Archer by Palimpsest Book Production Limited,
Falkirk, Stirlingshire

Printed in Great Britain by CPI Group (UK) Ltd, Croydon CR0 4YY

*As ever, thanks go to Sally and Thomas
for their love, support and patience*

INTRODUCTION

It's been my favourite joke of the year. Back in the day, a dog walks into a telegraph office in New York. He wants to send a message to his cousin in California. The operator says, 'It'll cost you two dollars.' The dog hands over his money and dictates his message. 'Woof woof. Woof. Woof, woof, woof. Woof.' 'For what you've paid, you can have two more woofs,' says the operator. 'But then it would make no sense,' replies the dog.

It's that mixture of logic and absurdity that underpins most of what appears in this book. Facts can simply be facts, but for them to be memorable, mind-boggling, thought-provoking or simply wonderful, they must have a certain *je ne sais quoi*.

In an earlier life, I used to edit a column in the weekly magazine *New Scientist*. Readers wrote in with clever questions and other, even cleverer, readers attempted to answer them. I just sat in the middle trying to figure out which of the answers was the right one. And, most importantly, which of the questions to ask in the first place.

One evening, after perhaps a glass of wine too many,

we tried to determine what would be the perfect question for the column. It had to be quirky or amusing. It had to be clever without being overly intellectual. But most important of all, we had to be sure that out there, somewhere, there would be an answer.

After much debate, we figured we'd found it, the question that fitted all the criteria. It was 'How fat do you need to be, to be bulletproof?' And yes, somebody really had done the research. The US military had the answer – you needed 60 centimetres of fat around all your vital organs in order to become bulletproof. The drawback? You'd probably be dead of a heart attack before you could even waddle into battle. Nonetheless, we felt smug – our perfect question had an answer.

Perhaps not all the facts in this book are as gratifyingly entertaining as that one, but you'll find the surprising, the witty, the ridiculous and the diverting. And perhaps one or two of them will encourage you to seek out more (beware, it's exactly what happened to me all those years ago!).

So, on to my second favourite gag of the year. It's a dad joke. Two men are convicted of stealing a calendar. They both got six months. This book, I hope, will appeal to dads, to mums, to their offspring, to their offspring's grandparents and pretty much to everybody. And, after reading it, you might be inspired to come up with a question that's even better than how fat do you need to be, to be bulletproof? If so, I'd like to hear from you.

Mick O'Hare

CHAPTER 1.

OUR BODIES

(or Yawns freeze your brain)

What? Yawns freeze your brain?

Sort of, but perhaps not in the way you think. There is no ice involved but some scientists believe yawning makes your brain cooler. This helps keep it at its optimum functional temperature – heat is released as we yawn – and studies show people yawn more in summer. However, there is no actual scientific consensus on yawning. A more popular theory than the brain-cooling suggestion – and the only one that takes into account the fact that we yawn most when we are sleepy – is that yawning helps keep us awake when we would otherwise have nodded off during boring or passive activities. It stimulates the carotid artery in the neck, increases both heart rate and blood flow to the brain and releases hormones that keep us awake. It also dampens our eyes, which stimulates us to wipe them. Because we rarely yawn when on the move, this suggests yawning is only necessary when we are passive. However, other researchers have noted that we will yawn before undertaking strenuous exercise or before an exam, when blood flow to the brain is equally desirable. Lastly, yawning helps reduce pressure and blockages in the Eustachian tubes in our ears. But this might just be an unintended consequence, because swallowing is a more effective means of doing this.

Why is yawning contagious?

There seems to be more agreement on this. In primitive societies, where it made sense for most people in groups to sleep at the same time and disturbances were kept to a minimum, yawning indicates that sleep time is approaching. This is backed up by the fact that we yawn if we see a family member or friend do it, but more rarely if we see a stranger yawn. The empathy of yawning draws social groups closer together and synchronises the collective mood. Other animals yawn too but contagious yawns are only found in more social animals, such as chimpanzees and lions, which gives weight to the empathy/synchronising theory. Even more interesting is that some animals, including dogs and elephants, are prone to catching yawns from humans. Chimpanzees will even yawn when they see a robot yawning. We are still figuring out why.

The wonderfully named Sprague-Dawley rat yawns about 20 times an hour, as opposed to only about twice an hour in other rat species. Studies show that the rat's facial temperature falls every time it yawns, allowing it to cool its body in hot weather.

Why do we have eyebrows?

They serve a very practical function by stopping sweat running into our eyes and stinging them when we are exercising or on hot days. But eyebrows serve another purpose: we use them to express our emotions, and it's why the hair of human eyebrows is usually noticeably different to our skin colour – to make them stand out. Surprise, annoyance, happiness, confusion, and so much more, can be displayed using our incredibly mobile eyebrows. For our ancestors, needing to know from a distance whether an individual or a group had hostile intentions or otherwise was an important evolutionary benefit. Anthropologists refer to the 'eyebrow flash' – a rapid up-and-down movement that also opens the eyes and conveys recognition and acceptance.

Tests in laboratory conditions during which speech is forbidden have shown that people with no eyebrows, or those whose eyebrows are immobilised, have greater difficulty conveying their emotions. Think twice before getting those Botox injections.

Why do we have fingerprints?

So the police can solve crimes. Well, it helps, of course, but obviously that was never their primary purpose. Fingerprints work much like the tread on a car tyre, helping us to grip things. Smooth surfaces are great for this in dry environments but are useless in wet ones. Our fingers have evolved to have raised and depressed areas that channel the water away from our finger ends, allowing us to hold onto things that might otherwise slip from our grasp.

Myth: Your fingerprints are unique.

It would appear that's not the case, at least not between each person's individual fingers. It's long been believed that the prints on all ten of our fingers are as unique as they would be if each finger belonged to a different person. But recent research shows that might not be true. In 2023, Columbia University ran a study using artificial intelligence which, when presented with two fingerprints, correctly predicted whether they came from the same person or not 77% of the time. Fingerprints from different fingers of the same person shared strong similarities.

Toe prints are as unique as fingerprints.

Nipples also leave unique patterns in the way fingerprints do. However, most criminals fail to remove their shirts at crime scenes so this knowledge, while intriguing, is of little use to the police.

How many atoms are there in a human body?

An astonishing 7,000,000,000,000,000,000,000,000,000 or thereabouts (7 octillion).

Atoms contain a lot of space. If all this space was removed our bodies would fit into a cube about 1/500 of a centimetre on each side.

Every atom in your body is billions of years old. And we really are made of stardust. Some atoms, such as hydrogen, were created in the Big Bang 13.8 billion years ago. Other, heavier, elements, such as oxygen, were created in early stars. When these stars exploded, they created even heavier elements, such as iron and magnesium. Because our body is composed of all of these and more, we are all pretty ancient.

There are around 20 elements that make up our bodies – of these 12 weigh in total more than four grams each. In the average human male, they are broken down like this:

Oxygen 52 kilograms
Carbon 14.4 kilograms
Hydrogen 8 kilograms*
Nitrogen 2.4 kilograms
Calcium 1.12 kilograms
Phosphorus 880 grams
Sulphur 200 grams

Potassium 200 grams
Chlorine 120 grams
Sodium 120 grams
Magnesium 40 grams
Iron 4.8 grams

(*Most of our body's atoms are hydrogen, but hydrogen is very light, which is why it's only third on the list.)

Why do we cry?

It's not just to clog up your nose, cause your make-up to run and embarrass you in public. It seems there are a number of reasons why humans cry and one of the most important, counterintuitively, is to make us feel better. Crying releases oxytocin (sometimes known as the love hormone, which plays a role in social bonding) and endorphins, the feel-good chemical messenger, which help to relieve emotional anguish, along with physical pain. That's why we cry when sad or distressed too. There are also social reasons why we cry. It indicates we require help or support, and showing vulnerability elicits sympathy and compassion. Similarly, we also cry to express our own sympathy. If a friend has suffered a bereavement, we may find ourselves crying alongside them. Look at how many people cry at funerals, not all of them closely related to the deceased. Of course, not all crying is a response to pain or distress. We can cry when we are happy or in love or by watching a child achieve a goal. Researchers believe that crying helps us process and categorise our emotions, whatever they may be.

Why do we laugh?

Because it's preferable to crying? Well, often it is, but neither seem to be a choice. We actually start laughing (and crying) before we can talk, so it obviously has a social function. Laughter signals to others that we are connecting with them. We are far more likely to laugh when we are in a group, especially during activities like watching comedy on TV, but less likely to do so alone. Interestingly, people speaking in social situations laugh more than those listening to them, hoping, apparently, that by doing so the listener will warm to them. Not only that, but in tests listeners could discern a difference between the laughter generated between strangers and that generated between friends. This shows that some laughter is forced in order to make a new person like you or to show you are no threat. It exhibits an offer of friendship. There are also physical benefits to laughter. It increases our oxygen intake and affects our heart rate and blood pressure (both up and down, which simulates exercise). Perhaps significantly, it releases endorphins – those feel-good chemicals again.

Why do we smile?

Much like laughter, smiling is a social signal, which communicates to people around us that our intentions are positive and unthreatening, and we are in good spirits. However, research has shown that sometimes smiling can be used as an attempt to cover up fear. Martial arts exponents who smile before a contest are more likely to lose. Other research has shown that genuine smiling involves wrinkling around the eyes which tends to be absent with feigned smiles. So don't forget to wrinkle next time the boss makes another terrible joke.

Why do we have goosebumps?

Goosebumps, or goosepimples, are a remnant from our evolutionary predecessors. They occur when tiny muscles around the base of each hair tense, pulling the hair more erect. This would have helped to keep us warm if we were still covered in fur, fluffing up our coats, making them better insulators.

We get a similar response when we are scared. This is because lots of mammals fluff up their fur when threatened, to look bigger and more menacing. Early, hairier humans used to have a similar defensive reaction. Unfortunately today we still get the sensation of hairs standing on end, but instead of looking aggressive we just look bumpy.

Why is hair different colours?

It's down to your genes, although a lot more research is required. However, more than ten genes seem to be involved, with one called MC1R being prominent. If this gene is active, your body will produce more melanin. Essentially, the more melanin that is present in the pigment cells of your hair follicles, the darker your hair will be. However, two particular types of melanin seem to influence hair colour – eumelanin gives people black or brown hair (or an almost total lack of it gives blond hair) while pheomelanin gives people red hair. It's the proportions that lead to differing hair types.

Some of us, your author included, have grey hair. Grey is actually the base colour of all hair. As humans age and cells die, these include the pigment cells in your follicles. As they die, the pigment is lost, and hair turns grey.

Myth: Your author has grey hair.

His wife has corrected him. Apparently, he has no hair.

Myth: You can go grey overnight.

Although there are many unsubstantiated cases of hair turning grey very suddenly, possibly due to stress (including, reputedly, the locks of Queen Marie Antoinette, following her capture during the French Revolution), research has shown that, in theory, it should be impossible. Hairs take longer than 24 hours to grow out and the process of a whole head of hair turning grey usually takes anything up to 20 years.

Why do we dream?

It's complex. Stuff like this always is. So we are going to try to make it simple. We do know the emotional centres of the brain seem to trigger dreams, rather than those centres involved with logic or reasoning. Dreams also seem to be somewhat autobiographical and are based on recent conversations you have had or things you have just done or thought about while awake. But while researchers can't agree definitively on the purpose of dreams, there are a few key theories. Dreaming may be a form of therapy, acting out emotional drama that you would otherwise suppress if you were awake. It may help you to sort through complicated thoughts and feelings. The lack of a logic filter also leads some researchers to conclude that dreaming releases our creative tendencies – artists often credit dreams with inspiring great works. How many times have you awoken with a great idea, or a solution to a problem? But numerous studies have shown that perhaps the most likely reason we dream is to help us store memories. Without sleep, memories can be lost very quickly and it's believed that dreaming about things helps to ensure those memories are stored more permanently with the unimportant ones discarded. As we said: complex stuff. In fact, it seems the whole field of dream research is a bit of a nightmare.

The oldest known study of dreams is *The Egyptian Dream Book*, a papyrus dating back to the reign of Ramesses II (1303–1213 BC). At that time it was believed dreams were prophecies. Any reported bad dreams were recorded in the book in red ink. And while we are talking colours, it seems that 12% of people dream only in black and white.

Myth: All our bones interlock as part of our skeletons.

Not so. The human neck has a bone that is totally separate from all the others, called the hyoid, or tongue-bone. It makes constructing skeletons for medical students very tricky because it always gets lost. Sing along now: 'The tongue-bone's not connected to the knee-bone...'

QUICK-FIRE FACTS

If laid out end to end, the blood vessels of one adult human would circle the Earth's equator four times (that's about 160,000 kilometres).

We blink on average about 518 million times in a lifetime. That works out to about 15 times a minute.

A blink lasts between 100 and 150 milliseconds.

We call muscles 'muscles' because the Romans thought flexed biceps looked like a little mouse sitting on your arm. 'Muscle' derives from the Latin for 'little mouse'.

Your body produces enough heat in 30 minutes to boil a litre of water.

Although the speed varies, human nerves can transfer information at about 400 kilometres an hour.

Your left lung is about 10% smaller than your right one.

And the reason your left lung is smaller is in order to accommodate your heart.

Humans breathe in about six litres of air per minute.

When you breathe in, you inhale around 25 sextillion molecules of gas (mainly oxygen and nitrogen). That's 25 with 21 zeroes behind it.

Your body uses only about a quarter of the oxygen you breathe in. You exhale the rest.

Your blood makes up about 8% of your body weight.

More than half of your bones (106 out of the 206 an adult possesses) are located in your hands, wrists, feet and ankles.

Only 2% of humans have green eyes. The largest concentrations are in Scotland and Ireland.

OUR BODIES

Eyes remain approximately the same size throughout your lifetime, which is why babies' eyes seem big and cute – they are disproportionately large.

The left testicle usually hangs lower than the right in right-handed men. The reverse is true for left-handers.

Statistically, a woman's left breast is larger than her right (65% to 35%).

Your small intestine is about seven metres long (or around four times your height).

The largest muscles in the human body are found in your bum. Your two buttocks are more properly known as your gluteus maximus.

About 70% of the dust in your home is dead human skin cells.

Our brains have no pain receptors.

The brain loses on average one gram of mass for each year of ageing.

20% of people have a bigger second toe than their so-called big toe.

Human head hair grows about 950 kilometres over the average lifetime.

YAWNS FREEZE YOUR BRAIN

A human eats about 32 tonnes of food over a lifetime and spends just over three and a half years eating it.

One in 18 people have a third nipple. It's called polythelia.

CHAPTER 2.

LIFE ON EARTH

(or Why is bird poo white?)

Why is bird poo white?

It's because birds poo and pee in the same package. This means the uric acid from their urine bleaches everything that is being excreted, hence their droppings being white. You'll have seen the bleaching effect of uric acid on your clothing if you've ever been hit by bird poo, although this fascinating chemical observation might not have been your first reaction as you got splattered.

How do dogs find their way home?

The longest journey a dog has made to find its way home is about 5000 kilometres. Bobbie was with his human family on holiday in Indiana in 1923 when he went missing. Six months later – the following year – he turned up at the family home in Oregon. Yup, that's 5000 kilometres. Bobbie was exceptional, but in 1973 a German shepherd dog called Barry got lost on holiday in Italy. Six months later, he turned up at his home in Solingen in Germany, nearly 2000 kilometres away. Biologists aren't entirely sure how dogs can navigate such distances but it's believed they can detect the Earth's magnetic field. In 2013, researchers in the Czech Republic and Germany had already concluded that dogs aligned their spines with the Earth's magnetic field when defecating. Who knows, maybe they just follow the, er, trail of evidence…

Even stranger than the stories of Bobbie and Barry, is that of Prince. In 1914, Private James Brown was serving on the Western Front in France during the First World War. Prince went missing from his West London home… and two months later turned up in Brown's trench. Nobody understands how he did it, but he was adopted as the regimental mascot and, like his owner, survived the war.

Magnetite is an iron-rich mineral, which will align with magnetic fields. And in 2023 it was found in the beaks of homing pigeons and migratory birds, leading researchers to believe that it is somehow used to direct the birds over long distances.

Snails also have a homing instinct. This was first noticed by British amateur gardener Ruth Brooks, who found that the snails she removed from her garden to protect her plants kept coming back. Researchers at Exeter University studied the phenomenon and discovered that snails would return to where they were removed from, unless they were taken more than ten metres away (or perhaps it just took them longer and nobody was there to see them arrive home).

Interestingly, when a group of Cornish snails (from the western tip of England) were taken to a garden in Hertfordshire in central England, they all lined up and headed west.

Which species is most at risk of extinction?

According to earth.org the Amur leopard, native to southeast Russia and north China, is the most at-risk species on the planet. There are probably fewer than 80 left. Poaching, killing for traditional medicine and climate change are all factors leading to their seemingly imminent demise.

It's not a rare animal but it is a rare event. In spring 2024 in the eastern United States, two species of cicada – one that emerges only once every 13 years and another that emerges only once every 17 years – surfaced from the soil where they spend most of their lives and began mating simultaneously. That's more than a trillion mating at once. It is the first time both species have appeared simultaneously since 1803 and it will not happen again until 2245.

What is the most abundant mammal in the world?

Humans. There are more than eight billion of us scuttling around the planet and that is increasing. We even beat rats into second place (there are around seven billion of those). Whether this is a good or a bad thing we'll let you decide. In a lowly third place are sheep, of which there are a mere 1.2 billion.

What is the most abundant animal in the world?

Nematode worms. There are 57 billion of them for every human and they make up 80% of all living animal species on Earth. They have a total biomass of about 275 million tonnes and are usually microscopic, although some rare parasitic types can grow to about one metre in length. They can live pretty much anywhere, including inside you.

What is the most abundant fish in the world?

The genus *Cyclothone* (or the bristlemouth). They are bioluminescent and inhabit waters deeper than 300 metres. It's believed there are a million billion of them and they are also the most abundant vertebrates on the planet.

Why are there freshwater fish and saltwater fish and why do they die if they swap environments?

It's all to do with osmosis. Osmosis is the process by which liquids cross semi-permeable membranes, such as the cells of animals. It allows molecules of a liquid to pass through the membrane from a less concentrated solution (such as the body of a saltwater fish) into a more concentrated one (such as seawater), in an attempt to equalise the concentrations in both liquids. Water will always move from a more dilute solution to a more concentrated solution. In fish it's called osmoregulation. Saltwater fish are, obviously, designed to survive in the salty sea. But because they are surrounded by a high salt concentration, water will be continuously 'pulled' out of their bodies by osmosis into the seawater in order to achieve equilibrium between their bodies and the surrounding sea. This is why saltwater fish continuously drink: to replenish the water that is constantly leaving their bodies. It's exactly the opposite with freshwater fish. Because freshwater fish have a higher salt concentration within them compared to their freshwater environment, water is continually entering their bodies via osmosis. This means freshwater fish continuously pee in order to remove this water from their bodies. Put them in the opposite environment and the saltwater fish will keep drinking although it no longer needs to. It fills its body with water until its cells are saturated, then they explode and it dies. Meanwhile the freshwater fish in seawater will keep peeing when it should be drinking and it will die of dehydration.

Why can some fish thrive in both environments?

The obvious example is the Atlantic salmon, which spends most of its life in the sea but returns to the freshwater river where it was born to spawn. Fish such as the salmon that can live in both environments are called euryhaline or anadromous species. There are far fewer of these than there are stenohaline fish (fish that can only live in either saltwater or fresh water). To do this, they actually have to change the chemistry of their own bodies depending on where they are. They have molecular pumps in their gill cells that siphon the sodium salts found in seawater in and out of their bodies. When in fresh water, they pump sodium in and when in saltwater, they pump sodium out. This way they can shift between their two different aquatic environments.

Myth: St Patrick banished snakes from Ireland

Except he didn't – Ireland's fossil record is entirely devoid of snakes, so it seems there were none for Patrick (who lived in the fifth century) to banish. The most recent ice age has kept the island too cold for reptiles and the surrounding seas have also kept snakes at bay.

New Zealand, too, is free of land snakes, for much the same reason as Ireland. However, it is home to two types of saltwater-dwelling snakes around its coastline. Anyone bringing snakes into New Zealand faces a heavy fine or a potential prison sentence.

How does a tiny chihuahua know that an enormous Great Dane is also a dog?

Or a spaniel or a dachshund or a poodle? They all look so different, yet they are exactly the same species (*Canis lupus familiaris*), and they can, in theory and in most cases, interbreed and produce viable offspring. The reason they look different despite all being the same species is that over centuries they have been artificially selectively bred that way by humans. Humans, too, look different depending on where they come from in the world, but this is by chance and the differences aren't so great. But it illustrates how the same species can appear outwardly dissimilar. So how do dogs, of which there are more than 400 breeds, know that other breeds are also dogs? It's mainly down to their impressive sense of smell which is 10,000 times better than a human's. Dogs secrete dog hormones and this is how dogs recognise each other despite being different breeds. But it's not only smell. Dogs shown pictures of other breeds alongside pictures of other animals such as cats and sheep, can correctly choose which ones are dogs.

Why don't birds have penises?

Well, actually some of them do, but 97% of them either have a very tiny penis or, in most cases, a cloaca. The cloaca is an exit in the male bird's body where the penis would normally be and it shoots sperm into the female's cloaca. However, it's all a bit, ummm, hit and miss, something which has, over the centuries, confused biologists because sperm delivered directly into a female using a penis leads to far more successful reproduction. They were doubly confused when scans of bird eggs showed that some male birds begin to develop a penis as an embryo but then it vanishes. Then in 2013 scientists discovered a gene called BMP4 which directs the penis cells to die back faster than they grow. As yet, nobody is sure why, but evolutionary biologist Patricia Brennan of the University of Massachusetts believes that it makes females more likely to engage in sexual activity. Mating in many bird species is forced by the male, but with cloaca, sperm cannot enter the female unless she cooperates. If she refuses, then sex simply doesn't happen.

Birds and reptiles evolved from a common ancestor, but studies of alligator eggs have shown that they lack the BMP4 gene. Sure enough, reptiles have kept their penises.

Of the birds that do have penises the bragging rights appear to go to the Argentine lake duck *Oxyura vittata*. Its coiled, corkscrew-shaped penis is longer than the bird itself. In 2001, one specimen's appendage was measured at 42.5 centimetres. Why evolution has gone to, er, such great lengths nobody is quite sure.

In mammals, penis size varies greatly. Surprisingly humans have bigger penises than gorillas (an average 13 centimetres to only 3.8 centimetres). A horse's penis is 45 centimetres while a rhino's tops out at more than a metre. But the blue whale takes home the prize with a willy longer than 2 metres.

LIFE ON EARTH

QUICK-FIRE FACTS

Around 10,000 new species of insect are discovered every year.

Three-quarters of all known animal species are insects.

Around 1 million insect species have been discovered so far but entomologists estimate we will discover as many as another 5 million.

There are probably more than a billion insects for each human on the planet.

All animals with complex nervous systems – such as mammals, birds, reptiles and fish – sleep.

Giraffes can sleep on their feet.

Giraffes are 30 times more likely to be struck by lightning than a human. Wonder why that is?

There are 0.003 lightning deaths per thousand giraffes every year.

Octopuses don't have tentacles, they have arms. Apparently, a tentacle has a sucker on the end. Octopuses' arms have suckers along their length.

Pigs were the first farm animals to be domesticated.

YAWNS FREEZE YOUR BRAIN

The world's oldest dog was Bluey, the Australian cattle dog, who lived to the age of 29 years and 6 months.

The oldest domestic cat was Creme Puff who lived in Austin, Texas and who made it to 38 years and 3 days.

All of humanity weighs six times as much as all the remaining wild mammals added together.

The oldest fossilised forest (possibly 385 million years old) was discovered in 2023 in a quarry north of New York.

A pelican's bill (better known as its throat pouch) can hold 11.5 litres of water or food, enough to feed it for a week.

Male snakes and lizards have two penises (known as hemipenes). Because snakes tend to mate in groups, it may be that being able to use the penis on the side of the nearest female is advantageous, suggests Christopher Friesen of the University of Sydney.

And, in the interests of equality and fairness, female snakes snakes have two clitorises (unsurprisingly known as hemiclitores).

Sadly for the tuatara reptile, it has no penis at all. Like birds, it has to rely on its cloaca.

LIFE ON EARTH

Gaboon vipers have the longest fangs of any snake – on average, a terrifying 55 millimetres. They also carry the most venom of any snake – around 600 milligrams.

Barnacles have the biggest penis to body size, 40 times longer than their bodies.

Pubic lice did not evolve from the human head louse. It evolved from lice that live on gorillas. How it made the leap to humans one can only, erm, speculate.

Clarinets are made from mpingo wood. It's very rare, which is why clarinets are expensive. Apparently, no other wood has the correct resonance.

Around 2000 new plant species are discovered every year, but two in every five are already threatened with extinction.

Approximately 27,000 trees are cut down every day.

Not all plants photosynthesise to survive. The fairy lantern plant which grows on the Japanese island of Kyushu has no chlorophyll and gets its energy from fungi that grow on it.

Every year plants photosynthesising convert about 200 billion tonnes of carbon dioxide into sugars (or 6000 tonnes per second).

In just two minutes, photosynthesis on Earth gathers the same amount of energy from the Sun as all the oil in the world's biggest supertanker.

Rice plants have about 15,000 more genes than humans do.

Bamboo can grow up to 90 centimetres a day.

CHAPTER 3.

THE UNIVERSE AND OUR SOLAR SYSTEM

(or How loud is the Sun?)

How loud is the Sun?

We are lucky the sound waves emitted by the Sun can't travel through the vacuum of space, because they would be incredibly loud if they could. The Sun is effectively a giant nuclear reactor, and nuclear reactors make a lot of noise. The Sun, as heard from Earth, would blast out the equivalent of 100 decibels, almost as loud as standing next to a speaker at a rock concert. And remember, that's from 150 million kilometres away. That said, presumably most fans of Motörhead or AC/DC would still sneer contemptuously.

Why does the Sun shine?

The Sun fuses hydrogen atoms in its core to make helium. And it fuses 600 million tonnes of hydrogen every second (the equivalent of igniting 100 billion tonnes of dynamite), which makes 596 tonnes of helium, while the remaining 4 million tonnes is converted into energy that we can see from Earth. Thankfully, the Sun still contains about 70% hydrogen, so we still have about 5 billion years of fuel left.

How loud was the Big Bang?

The Big Bang – the single, explosive event that led to the creation of our universe – was even louder than the Sun. But not by much. It probably peaked at between 110 and 120 decibels, which is painful to some human ears but probably still not enough to enthuse Motörhead or AC/DC aficionados. Cosmologists also reckon it wasn't a single earsplitting sound like an explosion, and was more likely to be a very loud hum. When John Cramer, a physicist at the University of Washington, decided to recreate the noise by using data a satellite had collected while investigating the electromagnetic radiation remnants from the Big Bang, he found the sound, although loud, was so low that its frequency had to be boosted billions of times for it to become audible to the human ear. It sounded "like a video game character dying", he said.

Can you still hear it?

Yes, you can. It was such a cataclysmic event that its echo – known as the cosmic microwave background – can still be detected. All you need is a radio receiver. Some of that fuzzy interference you hear as you tune it is the dwindling sound of the Big Bang.

What is a meteorite?

A meteorite is a meteor that has reached the Earth's surface.

Okay, so what's a meteor?

Most meteors are bits of asteroids, leftover pieces from the formation of the solar system. Asteroids usually sit orbiting the Sun in a belt of rubble between the planets Mars and Jupiter. Sometimes they collide, or get hit by other space objects, and fragments of them are set on a collision course with Earth, where usually they burn up in the atmosphere. But if they make it to the surface, then they are called meteorites. Some very special types of meteorite are actually pieces of Mars or the Moon, blasted off their surfaces by impacts from other meteorites.

How many meteorites do we know of?

Getting on for 35,000. The largest is a 34-tonne piece found in Greenland in 1894, which is on display at the American Museum of Natural History in New York.

Meteorites sometimes hit people. In 1954, Ann Elizabeth Hodges was sitting in her home in Alabama when a meteorite smashed through her roof, bounced off the floor and hit her, bruising her side. Ann, we can safely assume, learnt the difference between a meteor and a meteorite.

Why are planets spherical?

Because of gravity. A planet's gravity pulls the edges in equally from all sides. This makes the overall shape of a planet a sphere. All the planets in our solar system are spherical, however some are more spherical than others. Mercury and Venus are the roundest of all – nearly perfect spheres. But Saturn and Jupiter are a bit thicker around their middles. As they spin, they bulge along their equator. Gravity still holds the edges in but, as the planet spins, stuff is trying to spin out like mud flying off a tyre. And that stuff on the equator also has to move faster than things on the inside or at the poles to keep up. This is true for anything that spins, like a wheel or a fan. Things along the edge travel the farthest and fastest. And therefore they are thrown outwards. Hence the bulge.

So why aren't asteroids (spherical)?

They just aren't big enough. The bigger the body, the stronger the gravity. Planets form when material in space clumps together, but if not enough of it gathers, it won't form a strong gravitational field. This means the asteroid will keep its original shape.

What would happen if somebody stole the Moon?

Well, apart from having to call in whoever the galactic equivalent of Sherlock Holmes is, it would be very, very, very bad news. We'd lose the oceans' tides, which would have a terrible effect on our ecosystems. Plants that usually received nutrients twice a day would either be stranded high and dry or submerged, and all the animals that feed on them would suffer similarly. Nocturnal animals that use moonlight as guidance or to synchronise behaviour would be impaired. But the most cataclysmic effect would be that the Earth's orbit would become imbalanced – its position in the solar system is controlled by the mass of the Moon being so close. Our orbit would most likely become far more elliptical, leading to greater extremes of temperature and thus huge climate change. Not only that but the Earth's rotational axis is also controlled by the proximity of the Moon. We could all end up with six months of blazing sunshine followed by six months of icy frigidity. Looks like we should be beefing up our nearest celestial neighbour's security protocols.

The Moon is shrinking. Scientists have calculated that its diameter has shrunk by about 45 metres over several hundred million years. This is because the Moon still has a hot, molten core, just like the Earth, and as this cools it shrinks, meaning the outside of the Moon shrinks too, a bit like a wizened raisin. But don't try eating any of it, it doesn't taste like raisins or even green cheese. According to astronaut John Young, who tried eating some moondust (yes, really), it tasted of wet ashes.

Who owns the Moon?

Nobody. Although some private companies have attempted to sell plots there. And, even more strangely, in 1996 German Martin Juergens declared that the Moon belonged to him, it having been gifted to his family by Prussian King Frederick the Great in 1756. Most people ignored him. So the Moon belongs to nobody and that's because the 1967 Outer Space Treaty states that it is the 'province of all mankind'. To date, 109 nations have signed the treaty, which leaves approximately another 100 wondering if they can buy one of those plots before it's too late.

Another treaty signed in 1979, with a title so long most people never read to the end, introduced new laws about the Moon. The Agreement Governing the Activities of States on the Moon and Other Celestial Bodies bans the use of the Moon for military purposes and states that its resources are "the common heritage of mankind" and cannot be owned by states or organisations.

Myth: There is a dark side of the Moon.

'I'll see you on the dark side of the Moon,' sang Pink Floyd. Well, it seems they were mistaken. While it's true that one side of the Moon perpetually faces the Earth while the other – known as the dark side – faces outwards, both are exactly the same shade of grey. And both receive an equal amount of sunlight over the course of a lunar month (the time it takes – about 29.5 days – from a new Moon through a full Moon and back to a new Moon).

Myth: The Moon has no atmosphere.

Looks like we can fill an entire book of Moon myths, doesn't it? Although it is very, very thin, the Moon does have an atmosphere. It contains helium, argon, neon, ammonia, methane and carbon dioxide, held in place by the Moon's gravitational field. It also contains sodium and potassium, which are not usually found as gases in the atmospheres of Earth, Venus or Mars.

Why did humans go to the Moon?

In the 1960s the two world superpowers, the Soviet Union and the United States, were involved in a struggle to prove which political system was the best: Soviet one-party communism or American democratic capitalism. One of the ways they hoped to prove their case was by getting to the Moon before the other did. And while the Soviets put the first satellite and the first man and woman in space, the United States was the first nation to land a human on the Moon in July 1969. We also went to the Moon to study it but that was of secondary importance to politics.

Why haven't we been back to the Moon?

Mainly because there is no political imperative to do so. But also because, once it had been achieved, nobody wanted to commit vast amounts of money to doing it again. All the hardware needed to get there fell into disuse and then obsolescence, and the people who knew how to use it retired and died. Therefore, to go again we need to start from scratch. And that takes a long time, a lot of money and a lot of new technology. But at last, it is set to happen. America's space agency NASA has plans for a Moon landing in 2026 using its Artemis spacecraft. China also has tentative plans for a crewed landing.

Myth: We didn't go to the Moon.

We did. Here are some of the most outlandish theories as to why we didn't, debunked:

The American flag the astronauts placed on the Moon flaps, yet there is no atmosphere on the Moon to produce wind:

It doesn't. It's held aloft using a metal rod across the top, and it only moves when the astronauts are pushing it into the lunar surface. After that it remains still.

THE UNIVERSE AND OUR SOLAR SYSTEM

You can't see the stars in any of the photographs the astronauts took:

No, you can't. And if you could you might have a case for the photographs being fake. The reason you can't see the stars is because cameras need a long exposure time to pick up such tiny pinpricks of light. The photographs taken on the Moon were all well lit and had short exposure times, so that we could get clear, rather than blurred, photos of the astronauts and the vista. Go out at night with a camera and switch on any outside lights before taking a photo of the sky. The stars won't show up.

The shadows of the astronauts and the lunar module are all wrong:

Although most shadows on the Moon are caused by the Sun, light from the very reflective lunar surface and from electric lamps on the lunar module also cause shadows. All these different light sources from different angles mean shadows can form in many directions.

There is no dust on the lunar module's landing pads:

That's because the descent engine of the lunar module blasts it all away but, unlike on Earth, there is almost no atmosphere, so it doesn't swirl around, it just falls vertically to the ground, meaning none drifts back to rest on the pads. If there was dust on the pads, then, once again, you might have a case for fakery. But there isn't.

QUICK-FIRE FACTS

The universe started with the Big Bang. It may end in the Big Rip, when the expansion of the universe pulls its particles so far apart from each other that it simply tears. Cosmologists think that the earliest this can happen will be in 22 billion years' time.

Any star more than twice the size of our Sun will eventually collapse to become a black hole.

A black hole is an infinitely small, infinitely dense, region of space-time.

If the Earth collapsed into a black hole (which it can't) it would occupy a blob of 1.77 centimetres in diameter but would still weigh what it does today.

95% of the universe is invisible. We know it must be there but we can't see it and don't yet understand it. We call it dark matter and dark energy.

All the planets of the solar system can fit in the gap – when it's at its widest – between the Earth and the Moon (with 4392 kilometres to spare).

The Moon moves away from the Earth by about 3.8 centimetres every year.

THE UNIVERSE AND OUR SOLAR SYSTEM

Earth's Moon is the fifth largest in the solar system. Not bad out of 290.

If he was running at his fastest speed (43.99 kilometres per hour) it would take Olympic champion sprinter Usain Bolt 260 days to run around the distance of Saturn's rings.

Saturn's rings are 90% ice.

It's not just the planet Saturn that has rings, so do Jupiter, Neptune and Uranus, although they are much fainter.

Rings aren't limited to planets. In 2014 astronomers discovered rings around the asteroid Chariklo which is only 250 kilometres across. Unlike Saturn's, they are made of rock fragments.

There are more than half a million pieces of space junk orbiting the Earth.

Halley's Comet will return in 2061.

Mercury is the fastest planet in the solar system, travelling at 170,505 kilometres an hour, or 47 kilometres a second.

Neptune is the slowest planet, sluggishly making its way through the cosmos at 19,548 kilometres an hour, or 5.43 kilometres a second.

The presence of Neptune was predicted by mathematics before it was discovered.

Since Neptune was discovered in 1846, it has only just completed a full revolution (a Neptunian year) around the Sun (in 2011).

If all of the world's oceans were transported to the Moon, they would increase its diameter by 2% or about 70 kilometres.

As of February 2024, 644 people have travelled in space, according to the definition of space set out by the Fédération Aéronautique Internationale, the world governing body for air sports.

The United States Air Force has a different definition. They say 681 people have travelled in space. Either way, it's approximately 0.0000087% of the current population of the world.

The water on Earth was most likely carried here by comets that struck the planet about 4 billion years ago.

In 2023, the Irish bookmaker Paddy Power quoted odds of 200/1 for alien life being discovered before the end of 2024. Unsurprisingly, for conspiracy theorists at least, the US government itself was more confident saying there was about a 1 in 9 chance.

THE UNIVERSE AND OUR SOLAR SYSTEM

The same year, Paddy Power also offered 250/1 that alien life would be discovered before Tottenham Hotspur won a major football trophy following the departure of star striker Harry Kane to Bayern Munich.

It's estimated that there may be as much as 600 billion kilograms of ice in craters around the north and south poles of the Moon.

Only the United States and Luxembourg have granted their citizens the rights to mine and own the Moon's resources.

The United Nations' Outer Space Treaty is generally interpreted as barring individual countries from claiming ownership of lunar resources. The US and Luxembourg will need good lawyers.

CHAPTER 4.

YUCKY STUFF

(or Why do some people eat bogeys?)

Why do some people eat bogeys?

Some people eat earwax (see page 46) but bogeys are surely worse and yes, you'll probably have seen people eating them. Maybe you've even tried one yourself (the author is in denial mode here). But why? Obviously, bogeys are dried nasal mucus – a human produces between 1 and 2 litres of mucus a day. Most of it gets swallowed but some enters the nasal cavity, where it can dry out and crust, and produce bogeys. 91% of people admit to picking their nose, far fewer admit to eating the products (only 44%). But some people do and, amazingly, it seems there may be health benefits. Friedrich Bischinger, an Austrian doctor, reportedly tells parents to encourage bogey eating in their children. According to reports published by the United States National Institutes of Health, salivary mucins, which provide a protective cover around teeth and the inside of your mouth, are also present in bogeys and can, it seems, reduce dental cavities. These same salivary mucins also suppress certain microbes, such as the yeast *Candida albicans*, which causes the fungal infection known as oral thrush. So when your mum, or worse, your spouse tells you to stop picking your nose, remind them about the salivary mucins.

Myth: It's good to eat bogeys.

Never have we so quickly busted our own myth. Truth is, it's probably best to ignore what we said above. It seems the drawbacks probably outweigh the positives. Picking out all that dried mucus stops it from doing its job. It acts as a filter catching viruses and bacteria in the air you breathe. And nosebleeds are more common in repetitive nose-pickers, with infections commonly finding their way into your body through the lesions.

When we breathe, we favour one nostril. But this changes from one nostril to the other throughout the day, with the one not favouring breathing becoming more dominant at detecting smells. Why this is, we have yet to discover.

If, after reading all this, you remain a fan of eating bogeys, it's possible you've tried rheum (sometimes called 'sleep'), which you find around your eyes when you wake up. It's probably a bit better for you than bogeys, because although it's of similar composition, it's fresher and has fewer bacteria.

Why do humans use toilet paper?

Other animals don't need it, so why do we? It's partially an anatomical thing. Humans walk upright at all times, whereas other great apes don't and have a posture that sees their bottoms sticking out. So if, like them, we just let our poo and pee go as we walked around, it would run down our legs. Lovely! So instead we have to sit or squat in order to defecate and make sure our poo does not come into contact with our urinary tracts (which are situated very close together). We have also learnt that pathogens from our bottoms can easily infect our genitals, so wiping them away rather than leaving them hanging around is much healthier. And now that humans wear clothing, we also don't want to soil it (although we don't take risks and have invented underwear especially for that purpose).

In order to make it easier to test nappies, the Kimberly-Clark corporation in Dallas, Texas, invented fake human poo. All you had to do was add water to their mix of polyvinyls, starches, gelatins, gums and insoluble fibres and resins, and there before you was your synthetic turd. Nappy researchers the world over must have heaved (a sigh of relief).

Why is pee yellow?

All drinks pass through the kidneys and are converted into urine. Urine contains the yellow pigment urochrome. Sometimes we eat something, such as beetroot, which has so much pigment some manages to pass through our systems unchecked, making our urine noticeably pink. Try it yourself – it's also a fun way to get your children to eat a healthy root vegetable they'd otherwise turn their noses up at.

Why does cheese smell (sometimes of sweaty feet)?

Cheeses are made with four basic ingredients; milk, a starter culture of bacteria, rennet and salt. It's the bacteria used to make up the starter culture which influences the taste, texture and smell of the cheese. Cheeses, especially those with rinds, are perfect for microbes and so the bacteria along with yeast colonises the cheese rind resulting in a distinctive scent. The different bacteria used determine just how smelly your cheese will be as they break down proteins in the cheese and release pungent gases. For example, Limburger cheese uses a bacterium called *Brevibacterium linens*. And if you've smelt a ripe Limburger, you'll not be surprised to read that *Brevibacterium linens* is also responsible for smelly feet. It just loves the salty, sweaty environment between your toes.

Epoisses de Bourgogne is widely regarded as the world's smelliest cheese (yes, even smellier than Limburger). It is so effluvious that it is banned on public transport in France.

So why do we like eating this stuff?

Smells foul, tastes great. But how? Well, that's down to something called backward smelling. As we chew food it releases an aroma, which rises through the back of our mouths into our noses (see page 68). So you'd think it might still be an unpleasant experience, but this time the olfactory centres in your brain are combining the scent with the creamy feel of the cheese on your tongue, and the perception is changed. Now it tastes delicious (although, fair to say, there are some people who still think it's foul. More fool them).

Why do armpits smell?

It's not to warn others that the individual in question hasn't washed for a while, although it does fortuitously serve that purpose. The armpit has been described as 'the mother ship of body odour' and most of us know why. Moist, warm and dark, and containing sweat glands, it provides the ideal conditions for a bacterium known as *Staphylococcus hominis*, which, when it meets molecules of Cys-Gly-3M3SH (a cationic antimicrobial peptide, since you ask) found in human sweat, breaks them down into products called thioalcohols. When these evaporate, they smell variously of onions mixed with sulphur and sometimes rotting meat.

Ancient Egyptians were the first to realise that trimming armpit hair reduced armpit odour, while for centuries rubbing salt into your armpits has been popular in Asia. Modern science shows that both techniques hold true: hair harbours bacteria and salt kills bacteria.

Myth: Armpit smell is a turnoff

Well, it might be for some people. Maybe even most of us. But that oh-so-distinctive odour also contains pheromones, which are a sexual attractant. Online sales of, er, over-ripe T-shirts worn by both sexes are surprisingly lucrative. What may be a myth is that Napoleon Bonaparte, heading home from the battlefront, wrote to his lover Joséphine, saying, 'Don't wash, I will arrive in three days.' It seems he wasn't telling her to ignore the laundry. Many historians argue that masking bodily odours is something of a modern affectation – the first commercial deodorant didn't go on sale until 1888 (it was called Mum and is still a brand today).

Myth: Men and women need different types of deodorant

Well, you'd think so, wouldn't you, if you check out the shelves of your local pharmacy. As far as packaging, advertising and marketing goes, deodorants are, without doubt, delineated by their users. But, as long as you are not bothered by the fragrance of your deodorant or the colour of its packaging, you can buy any of them, they all do exactly the same thing – inhibit the growth of our old friend *Staphylococcus hominis*. However, marketing conquers all, even when it comes to armpit sprays, hence two separate aisles of pretty much identical products.

The average human gets through 544 underarm deodorants in a lifetime.

Of course, it's not just armpits and feet that smell. Researchers at the University of Pennsylvania discovered that subjects in blind tests could identify the sex of a person simply by their breath, with 95% accuracy. And that male breath, in general, is more pungent and less pleasant than female.

Why do we have earwax?

Earwax is produced by specialised sebaceous glands. It stops small fragments of dust or hair, but also microscopic bacteria and fungi, from entering the ear canal. These stick to the wax and are ejected as the wax is slowly pushed outwards. Some people like to aid this process using their fingers. We'll let readers decide how impolite this is but as a rule of, erm, thumb, the darker the wax is, the older it is, and thus it carries an increased likelihood of contagious contents. Some people choose to eat their earwax. Now you know why they shouldn't.

There are two types of earwax, wet and dry, and people have one or the other. Wet earwax is an orangey-brown colour and sticky. Dry earwax is scaly. Earwax is made up of dead skin in an oily secretion, plus water from sweat glands. Those with dry earwax lack the oily part because of a genetic mutation. Eat neither type.

Most people of African or European descent have wet earwax, while the mutation for dry earwax is what is known as 'recessive'. Unless both your parents have the mutation, you won't have dry earwax. But if you do, luck is on your side. The mutation also ensures what comes out of your sweat glands (see above) will be odourless. That's a lifetime's saving on deodorant.

Why is the Guinea worm considered the nastiest human parasite?

Because when you become infected by it, usually through drinking contaminated water containing its larvae, nasty things happen. First the adult forms inside your body and then it releases its own larvae. Blisters appear on your skin as the larvae attempt to escape your body. The blisters burn and burst, releasing the larvae into the outside environment. Then the adult decides it too wants to leave and starts to clamber out of one of the blisters. At this point you need to get hold of it and drag it clear. Not surprisingly, all this makes you very, very ill.

Myth: The Guinea worm is the nastiest human parasite.

Okay, so it depends on your opinion. And *Naegleria fowleri* (or the brain-eating amoeba) is much rarer than the Guinea worm, but it is especially nasty. The larvae are found in warm water and soil and enter humans through their noses. They find their way to your brain, where they start to eat away at the tissues. Within a week you'll almost certainly be dead. Makes Guinea worm blisters seem almost welcome.

Animals can fare even worse. *Cymothoa exigua*, or the tongue-eating sea louse, finds its way into the mouths of fish. There it feeds on blood from the fish's tongue until the organ collapses, whereupon it attaches itself to the tongue stump and feeds on blood and mucus in the fish's mouth.

Myth: Eating mummies is good for you.

Throughout history people believed that if you ate a piece of an Egyptian mummy, it could cure anything from bubonic plague to a sore toe. Medieval apothecaries used to sell mumia, a 'medicinal' substance created from mummified bodies. The desire to feast on mummies proved problematic for museums, which had to erect barriers around their Egyptian exhibits lest somebody turn up with a sharpened drinking straw and try to take a slurp. And while the Victorians had stopped eating mummies, the wealthy frequently bought or plundered examples from Egypt and, when removing the mummy's bandages, held 'unwrapping parties' in their own homes to show how rich they really were.

Is it ever good not to be yucky?

Sometimes. In the nineteenth century, Hungarian doctor Ignaz Semmelweis realised that if doctors washed their hands before entering maternity wards, there were far fewer deaths of mothers and babies. He changed medicine forever.

QUICK-FIRE FACTS

Men with chest and naval hair produce more belly-button fluff (see page 88) than men with less. The hair abrades the fibres of whatever shirt they are wearing.

A shirt worn 100 times contributes 0.1% of its volume to belly-button fluff, according to researcher Georg Steinhauser of Vienna University.

In 2009, a 55-year-old woman in Nebraska was suspected of having an abdominal tumour. When surgeons investigated, it turned out to be a 1-centimetre diameter ball of belly fluff.

The longest bout of constipation was 45 days until, in 2013, a woman in India had a faecal mass, the size of a football, removed from her intestine.

Two diseases, alkaptonuria and maple syrup urine disease, can be diagnosed by sniffing earwax.

Up until the start of the twentieth century, the blood of freshly executed prisoners was considered a health tonic and members of the public would pay executioners to drink it warm from the gallows.

The longest tapeworm removed from a human was 25 metres long.

Some religious groups consider that eating a baby's placenta is a form of cannibalism.

A twelfth-century text from China has recipes for making dumplings with minced human meat.

A kitchen sink contains more bacteria than a toilet, mainly because people regularly clean their toilets and neglect their sinks. But kids, don't wash your hands in the loo, okay?

Before we understood about germs and illness, people believed disease was caused by breathing in 'bad air', known as miasma.

Studies of mobile phones have discovered that one in six of them has faeces on its surface.

Around a fifth of office coffee mugs have traces of faecal bacteria on their inner surfaces.

YUCKY STUFF

Astronauts on long space missions 'redrink' their own purified urine.

A human produces around 100 litres of sweat every year, just while lying in bed. Sweat dreams...

The average human farts around 15 times a day.

One in five adults in the United States admit to urinating in the swimming pool.

Gastroenterologist Michael Levitt invented a 'Toot Trapper' – a cushion filled with activated charcoal, which absorbs gases, so he could analyse the constituent products of human farts.

In the United States canned fruit juices can contain as many as four liquidised maggots before the can's list of ingredients has to say 'maggots' on it.

Peanut butter can also contain five rat hairs before they too get added to the list.

The oldest head-louse egg to be discovered was found in an archaeological dig in Brazil. It was 10,000 years old and still attached to a human hair.

Every year about 15 million children in the United States will contract head lice.

A large proportion of the human population – perhaps as many as 50% – have microscopic mites (called *Demodex*) living in their eyelashes.

Mike the headless chicken, who lived in Fruita, Colorado in the 1940s, survived 18 months after having his head severed.

When two people kiss, they exchange between 10 million and 1 billion bacteria. But more bacteria are transferred through hand shaking.

CHAPTER 5.

THE WORLD AROUND US

(or Why is the Eiffel Tower taller in summer?)

Why is the Eiffel Tower taller in summer?

Because of thermal expansion. In summer it's hotter so the atoms in the tower's pylons become more agitated and take up more space, and so the tower grows 15 centimetres (making it 330 metres and 15 centimetres tall). The top can be seen from 59 kilometres away (or perhaps 59 kilometres and a bit more in summer). In 1971, 160-kilometre-an-hour winds made the tower sway a record 15 centimetres, although it can survive five times this.

It is illegal for professional photographers to take photographs of the Eiffel Tower at night. Even though the Eiffel Tower is legally a public space, the lights on its night-time display are not. They are the copyright of Pierre Bideau, the artist who installed them in 1985.

Myth: The Eiffel Tower is made of steel.

Most people believe that to be so but it's actually made of iron, which means it needs to be constantly painted to stop it rusting. The colour used is known as 'Eiffel Tower beige' in three different shades, which help to emphasise its height. It takes seven years to paint.

What is rust?

Rust is corroded (or corroding) iron. When oxygen in the air comes into contact with iron – especially the refined iron used to build bridges, ships and Eiffel Towers – in the presence of water, such as rain or the ocean, it corrodes into reddish-brown hydrated iron oxide. Or what we all call rust.

Myth: The Forth Railway Bridge near Edinburgh has to be constantly painted to stop it rusting away.

This aphorism has led to the phrase 'like painting the Forth Bridge' being applied to any repetitive task. And it's true that it was once the case. The painters would start at one end of the bridge, paint for three years until they reached the other end and then start all over again. But modern anti-rusting paint compounds mean that it's no longer the case. Since 2002 the bridge needs painting only once every 20 years.

Why do aeroplanes have windows with rounded edges?

The same reason ships have round portholes. If they had square corners, the constant flexing of the metal would begin to rip at these points. This is far less likely to happen with rounded corners because the pressure is more evenly distributed and a crack in the fuselage, or a broken window, less likely.

The deadliest air accident was on 27 March 1977, when two Boeing 747s collided on a runway in Tenerife in the Canary Islands and burst into flames; 583 people died.

The first person to die in an air accident was Jean-François Pilâtre de Rozier on 15 June 1785. His hot-air balloon burst and he fell from the sky near Calais. Two years earlier, on 21 November 1773 in Paris, he had made the very first un-tethered hot-air balloon flight alongside François Laurent d'Arlandes.

There were only 66 years between Orville Wright's first powered aircraft flight at Kitty Hawk in North Carolina (17 December 1903) and Neil Armstrong stepping foot on the Moon (02.56 GMT on 21 July 1969).

When the United States' space station Skylab came plummeting back to Earth in 1979, debris was scattered across a remote part of Western Australia. The NASA investigating team that arrived to pick up the bits and pieces were handed a $400 fine by the Australian authorities for littering. It took 30 years before the bill was settled by an American radio station embarrassed to hear that NASA had never paid up.

What's the largest city in the world?

Depends how you look at it. Defining the largest city in the world by land area is difficult. Large parts of Greenland are defined as municipalities. They cover hundreds of thousands of square kilometres but have very sparse populations. Using this definition, the largest city in the world is Sermersooq which covers 531,900 square kilometres. However, Sermersooq is not a city in the way most people would define one. So, in the more obvious sense, the largest city by area in the world is New York, with an area of 12,093 square kilometres. However, the biggest city in the world by population is Tokyo, with 37,115,035 residents in 2024. It's shrinking, though. The 2024 figure is down by about 80,000 on the previous year. It may soon be overtaken by Delhi.

Myth: The Statue of Liberty is in New York.

Ignoring the fact that the original is in Paris, the Statue of Liberty on Liberty Island in New York Harbor was officially just inside the state of New Jersey. However, in

1987 the Supreme Court of the United States deemed it to be in New York to avoid confusion, thereby confusing everybody.

What is the most protected building in the world?

The United States Bullion Depository, better known as Fort Knox. It holds half of the gold reserves of the US (8134 tonnes, worth around 300 billion dollars), has an 18-tonne blast-proof door, which is set into concrete-lined granite and reinforced steel and can withstand a nuclear attack. It is also home to 40,000 soldiers plus depository employees. The building has an escape tunnel if, for whatever reason, you find yourself locked inside (you might need to use it because, once locked, the door has a 100-hour time seal).

Meanwhile, the most protected place in Europe is probably the archive in Vatican City in Rome. The Roman Catholic Church possesses around 35,000 rare and sacred documents that are stored there. Permission to enter is rarely granted and, when it is, only one special research officer can enter at a time.

Myth: 10 Downing Street is the most protected building in the UK.

Actually, it might not be. And it might not be Buckingham Palace or the Bank of England either. Apparently, it's

Bold Lane Car Park in Derby. Yes, a car park. In Derby. The car park is multi-storey, with 440 parking bays, and is open 24 hours a day. Drivers are issued with a barcoded ticket. This code is scanned and linked to a specific parking bay once the driver types in the bay number. This activates a motion sensor in the bay beneath the car. When the driver returns, their barcode allows entrance to the car park and when they pay their parking charge the sensor is turned off. If they have not returned or they do not pay and the car moves, the sensor triggers an alarm. As well as detecting the horizontal motion of the car being driven off, the sensor also checks for vertical motion such as that caused by a person getting into the car. The car park's control-room operator can close any or all of the exits with barriers if security is breached and has 190 CCTV cameras and a computerised map of the building to follow the movement of any rogue car. There are also panic buttons located throughout the car park. The measures were introduced in the 1990s when the car park became a popular location for crime. Government cabinet meetings have not yet been held there, though.

Myth: A multi-storey car park in Derby is one of the most secure buildings in the world.

Surely it must be a myth? Surely? Contact us if you know different.

Myth: 1 horsepower is the energy output of 1 horse.

Scientist James Watt defined 1 horsepower as 'the amount of work required from a horse to pull 150 pounds out of a hole 220 feet deep'. He invented this unit of energy so he could sell his engines which he said could 'do the work of ten horses' (or 10 horsepower). But was he right? Well, only sort of. The average horse galloping at full speed pulling a coach or carrying a rider has a maximum output of 15 horsepower, which suggests Watt's calculations were a bit off. However, over the course of 24 hours the average power output of a horse is around 1 horsepower. So maybe he wasn't wrong but was instead guilty of trying to oversell his engines.

Ironically, the maximum output of a human is pretty much a single horsepower, although this can briefly be exceeded by elite athletes.

A Formula 1 car produces an output of up to 1000 brake horsepower. It's called brake horsepower in cars, rather than simply horsepower, because this takes into consideration the friction between a car's tyres and the road. The 'braking' effect and power loss caused by friction means a car's brake horsepower is always less than its actual horsepower.

YAWNS FREEZE YOUR BRAIN

QUICK-FIRE FACTS

The world's first oil well was in Titusville, Pennsylvania in 1859. The locals used the oil as medicine.

Five countries power their electricity grids with 100% renewable energy (wind, solar, hydro or geothermal): Costa Rica, Iceland, Norway, Paraguay and Uruguay.

Around 30% of global electricity comes from renewable forms.

Britain's Royal National Lifeboat Institution has saved more than 144,000 lives at sea since it was created in 1824.

Each of the five engines on the first stage of the Saturn V rocket which took astronauts to the Moon expelled 2,542 litres of gas a second.

The Wunderland Kalkar amusement park and hotel complex in North Rhine-Westphalia in Germany was originally intended to be a nuclear power plant.

Tropical Islands in Berlin – the largest indoor water park in the world – was formerly a hangar for Nazi airships.

There were more than 45,000 concrete blocks in the 155 kilometres of the Berlin Wall.

THE WORLD AROUND US

The world's shortest scheduled passenger flight lasts 57 seconds in good weather. The 2.7-kilometre flight is between Westray and Papa Westray in Scotland's Orkney Islands. The record for the journey is 53 seconds.

Six London Underground stations are named after pubs. Royal Oak, Angel, Swiss Cottage, Maida Vale, Manor House and Elephant & Castle.

There would have been a seventh named after a pub but the Bull & Bush station on Hampstead Heath was never opened, although the platforms were built.

Discounting microstates, the only mainland European capital city without a railway station is Tirana, Albania.

45% of Americans have no access to public transport, defined as a public transport stop within 500 metres' walking distance.

Toilets were first introduced on British trains in 1873, but only in sleeper carriages. Other travellers just had to hang on...

In 1910, kissing was banned on French railways because it could cause delays. It's not clear who was kissing who and why that would stop a train, but it was definitely banned.

The world's longest and heaviest train ran on 21 June 2001 between Newman and Port Headland in Western

Australia. It weighed 90,477 tonnes and was 7.35 kilometres long.

The first sitting United States president to travel by train was Andrew Jackson in 1833.

In 1971, Porsche designed a racing car that the team's sponsors Martini & Rossi decided looked like a pig and was so ugly they refused to let their logos appear on it. At the last minute, Porsche designer Anatole Lapine painted it pink and labelled each of the body parts according to butcher-style pork cuts. Martini & Rossi regretted their decision – the Pink Pig Porsche was adored and still appears in photographs today.

Brooklands in Surrey, in southern England, was the world's first purpose-built motor racing circuit. It was opened in 1907 because the British government had banned motor sport on public roads.

Briton Lewis Hamilton has won more Formula 1 Grands Prix than any other driver (105). But the most successful driver by percentage of wins to race starts is Argentinian driver Juan-Manuel Fangio who won 47.06% of the Grands Prix he started compared to Hamilton's 30.35 (as of August 2024).

The most popular road car colour is white.

Concrete is the world's most used building material.

The largest concrete structure in the world is the Grand Coulee Dam in Washington state, USA, which used up 9,155,943 cubic metres.

The tallest dam in the world is the Jinping-I Dam, at 305 metres high.

The world's widest dam is the Kariba Dam on the Zambezi River between Zambia and Zimbabwe. It is 579 metres long.

The first recorded architect in history is Imhotep, who designed the Step Pyramid at Saqqara and became chief architect for Pharaoh Djoser of the Third Egyptian Dynasty.

Architecture used to be an Olympic sport (as were painting, literature, music and sculpture). They were removed from the Olympic programme in 1949.

The longest time taken to complete a marathon was 54 years, 249 days, 5 hours, 32 minutes and 20.3 seconds, by Shizo Kanakuri of Japan. He started the Stockholm Olympic Marathon in 1912 but collapsed while running it, returned home to Japan, and was invited back to Sweden in 1967 to complete the race.

CHAPTER 6.

EATING AND DRINKING

(or Why do onions make us cry?)

Why do onions make us cry?

Tears are an onion's self-defence mechanism. When an onion is cut, it releases a volatile gas called Propanethial S-oxide, which makes your eyes tear up. This deters animals from eating or damaging the bulb. The good news is that varieties now exist that promise not to make you cry. Propanethial S-oxide is derived from amino acid sulphoxides, so the onions are bred without these. You can find them in your local supermarket but beware, avoiding tears will cost you about 50% more than your standard onion (and, apparently, they taste less oniony).

Why do we cook food?

It's a ubiquitous human activity. People all over the world, and dating back as far as our ancestor, *Homo erectus*, a million years ago, realised that heating food changed it. It made it easier to digest and crucially killed off any bacteria it was harbouring. However, they also discovered it tasted better, which is down to something called the Maillard reaction, named after a French chemist. It's a reaction between sugars and amino acids that happens when foodstuffs are heated and it creates those brown colours you see when meat is roasted or bread toasted. Interestingly, when given a choice, great apes such as gorillas and chimpanzees also prefer food that has undergone the Maillard reaction, even though they don't cook their food.

Potatoes are 98% digestible when cooked, as opposed to only 32% digestible when raw.

Myth: Spaghetti bolognese is Italian.

The meat sauce from the north Italian city of Bologna is known in Bologna as ragù. And it's eaten with fresh egg pasta, usually tagliatelle. Spaghetti bolognese, meanwhile, developed after Italian immigrants began to move to the UK and the United States around the start of the twentieth century. They had to adapt the dish to what they could find, using dried pasta, while the locals added in things that they thought sounded Italian such as garlic, herbs and mushrooms, none of which are found in ragù.

It's estimated that approximately 500 million family meals of spag bol are cooked in the UK every year.

Myth: Spaghetti grows on trees.

Okay, nobody believes it now. But because pasta was seen as such an exotic food, in 1957 the BBC show *Panorama* managed to convince a large swathe of the British population that the spaghetti harvest was under way in Italy. The programme was broadcast on 1 April.

Myth: Fish and chips are the quintessential British food.

Well, they are seen as such today. But it's most likely that fried fish in batter was first eaten in London by Jewish immigrants from Spain and Portugal from around the 1600s. Fried potatoes, or 'chips', were added later, possibly originating in Belgium or the Netherlands, and the first record of shops selling both comes from the 1860s. However, Charles Dickens makes the first literary mention of fish-and-chip shops in *A Tale of Two Cities* in 1859.

British prime minister Winston Churchill insisted that fish and chips should not be subject to rationing during the Second World War. And they weren't. Because he was the boss. (He also somehow seemed to circumvent the rationing rules on champagne, whisky, brandy, cigars...)

Myth: A human consumes around 2,000 kilocalories a day.

Way back in hunter-gatherer times that might have been true, because the only energy a human consumed back then was solely from the food they ate. But now a human in the western hemisphere consumes 100 times that, mainly through using coal, oil and gas for transport, heating, cooking, electricity and making consumer goods.

Myth: Fresh fruit and vegetables are better than frozen.

In fact, the opposite may be true. Nutrients are lost from the moment fruit and veg are picked and freezing quickly helps to preserve these. They last longer too.

Myth: Fat is bad for you.

The *Journal of the American Medical Association*, following an eight-year study of 49,000 women, found that low-fat diets offered no noticeable health benefits. The famed Mediterranean diet, which derives as much as 40% of its calorific value from fat, is considered extremely healthy. However, the fats in this diet are unsaturated and come from olive oil, fish and seeds. You still need to avoid the kind of saturated fats found in butter, cheese, cream, fatty meat and biscuits, which are linked to heart disease.

Why do we learn to like some foods as we age?

So many of us find foodstuffs such as olives, coffee without sugar, blue cheese and whisky unpleasant the first time we try them. But as we get older, we grow to like them. That's because our brains initially react to bitter or sour tastes by presuming that they are spoiled or poisonous. Eventually we learn that they are safe to eat and instead of the kneejerk reaction to spit them out, learn to enjoy their unique tastes. There is also social learning and peer pressure – we see adults eating them, or our contemporaries, and want to appear similarly grown-up and sophisticated, because things like whisky and olives are considered grown-up fare. Some people say that you need to try a food fourteen times to be sure you *really* don't like it. This has yet to be proven.

Myth: We taste using our tastebuds.

Well, only sort of. True taste from the tongue's tastebuds comprises only salt, sweet, bitter, sour and umami (the savoury taste). Most of what we perceive as taste is actually flavour, a process almost identical to smelling that is generated by the olfactory receptors in the nose as food is chewed and its odour rises through the back of the nose, where it is detected as coffee, strawberries, cinnamon or chicken. In fact, we effectively taste with our brains – all that information from our tongues and olfactory receptors is gathered by our brain and turned into the sensations we experience when eating.

EATING AND DRINKING

Myth: Tastebuds are found on your tongue.

Yes, that's true. But they are also found on the top of your mouth and in your throat. And remember that map of the tongue you might have been shown at school? The one which showed which regions of the tongue detected the different tastes? Thoroughly discredited. All regions of the tongue detect all tastes, though some are more sensitive than others. There's more. Those bumps on your tongue aren't tastebuds, they are fungiform papillae (or, er, mushroom-shaped nipples if you don't speak Latin). Your tastebuds reside inside and around them.

Myth: London dry gin has to be made in London.

Actually, it doesn't. London dry gin, beloved around the world because of the way it pairs so perfectly with tonic water, can be made anywhere. London dry is a style, not a controlled designation of geographical status, such as champagne or feta cheese. To qualify as a London dry gin, the spirit must contain juniper and have no artificial ingredients and can be watered down to a minimum of 37.5% alcohol by volume. It must also not be sweetened in excess of 0.1 grams of sugar per litre – hence its 'dry' designation. However, Londoners would argue, and with some justification, that the best London dry gins are still produced in the city.

Was James Bond an alcoholic?

In the previous book in this series we learnt why James Bond ordered his martinis shaken, not stirred. But we didn't address the trickier question of whether Bond drank too much. In 2013, the *British Medical Journal* calculated his alcohol intake. It would have floored an elephant. Here goes: on an average day, Bond consumes five vodka martinis. But that's only an average. In *From Russia with Love* he downs 50 units of alcohol over one 24-hour period, while in *Quantum of Solace* he knocks back six martinis on a single transatlantic flight. In the films he orders a drink, on average, once every 11 minutes. The *BMJ* researchers decided Bond was a 'problem drinker' who should be 'referred for further assessment'. But, some fans argue, one more reason he had his martinis shaken with ice was to make them more dilute. He doesn't drink as much as it appears. After all, he has to shoot straight. And plenty of people have pointed out how often Bond orders a martini in the films and then never drinks it. And, last of all, let's not forget, James Bond is fictional. No, really, he is.

Why do drinks such as ouzo or sambuca turn white when you add water to them?

These are anisette-based drinks (they taste of aniseed) and they contain aromatic compounds called terpenes. Terpenes are soluble in alcohol but not in water. So when you add water or ice, the drink becomes cloudy.

EATING AND DRINKING

Myth: Eating carrots can't turn you orange.

Oh, yes, it can. You have to eat a lot, but carrots are rich in a pigment known as beta-carotene. In humans, this pigment is usually converted to vitamin A by cells in the small intestine. But when high levels of beta-carotene are consumed, not all the pigment is converted and some of it circulates in the bloodstream. If these high levels are sustained for some time, the skin – especially the nose – may take on an orange hue, a condition known as carotenemia. It remains unknown whether former United States president Donald Trump is a fan of carrots.

Myth: Cheese gives you bad dreams

To debunk this myth, the British Cheese Board carried out a study in 2005. Of the 200 participants who ate 20 grams of cheese half an hour before bedtime, two-thirds reported remembering their dreams, with none of them reporting nightmares. We can only hope there are no holes in the study.

What foods cause cancer?

Earlier this century a cancer researcher with an eye for a joke set up a blog called the *Daily Mail* Oncological Ontology Project, because it seemed barely a week would go by without the British newspaper declaring that a certain food could cause (or in some cases cure) cancer. Most of the stories were, at best, subjective. However, verifiable research by oncologists has shown that the following foods can – although not in all cases – cause cancer: red meat (especially chargrilled), processed meats (such as bacon and salami), processed foods (such as sugary cereals or salty crisps), salt, sugar and sugary drinks, saturated fat (such as butter, palm oil and cheese) and alcohol.

Myth: There are foods that prevent cancer.

No single food can prevent cancer but by eating a balanced and nutritious diet rich in fruit, vegetables, grains, fish, chicken and olive oil, and avoiding eating too much salt, sugar, fat and the cancer-causing foods above, you are less likely to contract the disease.

If this didn't cause cancer, it would be a surprise. In 2002, the United States military invented the indestructible sandwich. It was designed to stay edible for three years at 26 degrees Celsius (or if conditions were extreme, six months at 38 degrees Celsius). The sandwiches were treated with humectants, which stop liquids leaking from the ingredients, and sealed in oxygen-free pouches to stop bacterial growth with sachets of oxygen-scavenging chemicals. Yummy.

Why do some religions have dietary restrictions?

Jews don't eat pork; Hindus don't eat beef. And for some Christians it's fish on Friday. But why? Some historians believe that God or various deities decreed this and that will suffice as a reason. Others think that the restrictions came about as a way to ensure people didn't get ill from food-borne pathogens or rotting produce. For example, pork needs to be thoroughly cooked and in the past that wasn't always possible, or in hot countries storage of certain items such as shellfish was difficult. Other religions believe in non-violence, which also extends to animals, therefore vegetarianism is common in Hinduism and Buddhism. Some religions, of course, don't allow alcohol, because it damages your body and also changes your relationship with your particular deity. The author is not a member of one of these.

Rastafarians are only allowed to eat fish if they are no longer than 12 inches (30.48 centimetres).

QUICK-FIRE FACTS

The fear of cooking is called mageirocophobia.

Strawberries are the only fruit with seeds on the outside.

The biggest blueberry ever grown was picked in November 2023 in Corindi, New South Wales. At 39.31 millimetres in diameter, it was the size of a golf ball and weighed 20.4 grams, ten times as big as the average berry. It went into a pie.

YAWNS FREEZE YOUR BRAIN

The largest tomato, grown by Del and Julie Faust of Minnesota, weighed 5.284 kilograms and had a circumference of 82.55 centimetres. An average tomato weighs about 0.14 of a kilogram.

Germans eat the most sausages. In 2020, they ate 1.5 million tonnes, or 27% of the world's output. Poland was second with 574,000 tonnes, followed by France on 495,000 tonnes.

Bananas were first displayed in a shop window in London in 1633. Today more than 5 billion are eaten in Britain every year (around 100 per person).

Around 165 million teabags are used in the UK every day.

Turks drink more tea than the British (almost 3.2 kilograms per person per year).

Enough Nutella is sold every year that it could cover the Great Wall of China (which by some estimates is 21,000 kilometres long) eight times.

The pecan is the state nut of Alabama.

The most expensive bottle of alcohol ever sold was a Macallan 1926 single malt whisky, which fetched £2.1 million at auction in November 2023.

EATING AND DRINKING

British chemist Dr David Clutton is the only person in the world to have a PhD in gin. Who marked his paper?

In Russia, until 2011, any drink under 10% alcohol by volume – and that includes beer – was classed as a soft drink.

The oldest alcoholic drink was made in Jiahu, China 9,000 years ago by farmers, using fruit and rice, which they fermented with the sugar from honey.

The first wine as we would know it, fermented from grapes, was made by villagers in the Zagros mountain range, which spans Iran, Iraq and Turkey. They added tree resin to preserve the wine, making it taste like white spirit. Lovely.

Romanian men are the heaviest alcohol drinkers on the planet, according to a 2024 survey. The average male consumes 27.3 litres of pure alcohol a year.

Caesar salad isn't from Italy. Nor did Romans eat it. It was invented in Mexico in 1927 by Caesar Cardini at his hotel in Tijuana.

During the American War of Independence, Americans avoided eating sandwiches because they originated in Britain.

Spam is short for spiced ham (and also, of course, for emails you really don't want – see page 113).

YAWNS FREEZE YOUR BRAIN

Hawaii consumes more Spam per capita than anywhere else.

There is more water in a cucumber (95%) than there is in a watermelon (92%).

Nutmeg contains the hallucinogen myristicin. Myristicin is also an insecticide.

American cheese was invented in Switzerland by Waltz Gerber and Fritz Stettler in 1911 when they were looking at ways to increase the shelf-life of dairy products.

However, the first commercially available slice of American cheese *was* produced by an American, James L. Kraft.

American cheese slices contain, on average, 51% cheese.

The Mexican dish Chimichanga was invented in Arizona. Chimichanga (almost) means 'thingamajig' in Spanish.

McDonald's sells 75 hamburgers every second.

Pork is the most widely consumed meat in the world, accounting for 36% of all meat sales.

Bananas float in water.

CHAPTER 7.

HEALTH AND ILLNESS

(or Why does silver foil make your tooth fillings hurt?)

Why does silver foil make your tooth fillings hurt?

If two different metals (the aluminium foil and the amalgam in your tooth filling) are separated by an electrolyte containing salts (your spit) an electric current will flow between them. And because the amalgam is close to the nerve in your tooth, that's going to hurt. If you are brave (or foolish) enough to try it, put a silver teaspoon in your mouth and touch a filling with it. Ouch!

Myth: Sugar causes cavities in your teeth.

It's actually bacteria – mainly *Streptococcus mutans* – attracted by the sugar that cause the cavities. The bacteria produce an acid when they feed on the sugar and the acid eats away at your tooth enamel. Sugar might be bad for you in many other ways but if you have something sweet, brushing it away will reduce the build-up of bacteria.

What is the rarest illness?

According to the *Journal of Molecular Medicine*, RPI deficiency is considered to be the rarest disease in the world. Ribose-5 phosphate isomerase (RPI) is a crucial enzyme in a metabolic process in the human body. Without it a sufferer will experience muscle stiffness, seizures and reduction of white matter in the brain. The only ever case was diagnosed in 1984.

Other rare illnesses include Fields' disease, a neuromuscular disorder, only ever diagnosed in one pair of twin sisters; aquagenic urticaria (or allergy to water), in which sufferers get red, itchy skin and can also be allergic to snow and rain; and kuru disease, which is only experienced by the Kuru tribe in Papua New Guinea, owing to their intriguing practice of consuming family members – specifically their brains, which contain an infectious protein – after their death.

What's the most common illness?

The common cold, the clue is in the name. Adults typically have two to three infections a year, while children may have as many as 12. There are more than 200 virus strains causing colds and these constantly mutate, meaning it's almost impossible to create a cold vaccine. We just have to suffer in silence (or not, if you are a man. Obviously).

What's the deadliest disease?

Although the cold may be the most common illness, it's not the most deadly. Heart disease kills more people than any other form of illness, with coronary artery disease the biggest killer among all forms of the condition. Covid briefly overtook heart disease as the world's biggest killer during the pandemic of the early 2020s but heart disease is now back on top of the list. Heart disease kills more people every year than cancer, war, terrorism, hunger, suicide, diabetes, respiratory diseases and mental disorders combined.

The stomach bug norovirus is considered the disease that is most easily transmissible, although of all the contagious diseases tuberculosis remains the deadliest, according to the World Health Organization, despite the fact that it can be cured. It is followed – in non-pandemic years – by HIV/AIDS and malaria.

When was the first brain surgery?

Amazingly, it was in the Stone Age (so that's at least prior to 3,300 BC) and was probably carried out on people who were in a coma, in an attempt to revive them, meaning that Stone Age humans were aware that the brain was important to human consciousness. We have evidence in the form of grooves carved into skulls, holes drilled into skulls and portions of removed skull. What's more, in many cases patients survived, because we can see that new bone growth had formed, which would only be the case if patients were alive.

Why do faeces vary in colour?

Faeces are normally brown-coloured because of the presence of stercobilin, a by-product of broken-down blood cells, and bile, which we use to digest fat. However, you'll probably have noticed some variation in that colour. And, thankfully, it's usually to do with your diet. People who eat lots of leafy vegetables, such as spinach or cabbage, might notice a green tinge to their poo. Tomatoes, red peppers and especially beetroot will colour poo red (worried borscht eaters in Russia routinely pester their doctors if their faeces are the boring brown the rest of the world is used to). And eating too many carrots really can turn your poo orange. However, if your poo remains constantly an unusual colour or if it is consistently black, white or pale yellow, it might be worth a trip to the GP.

According to gastroenterologists, we should, apparently, take between 10 and 15 minutes to pass a stool. Which actually has the author rather worried because, unless there is an especially interesting story in the newspaper, he usually spends no more than a minute or so atop the lavatory.

How many toilet rolls does a human get through in a lifetime?

About 650,000 apparently – or 20,000 sheets a year – but who's counting them? We also get through 35 tonnes of food, 780 pairs of socks and 15 toasters. (For the record, the author is only on his second toaster, he is clearly a careful owner. Or has his toast cooked very lightly.)

The first toilet paper was invented by the Chinese, shortly after paper was invented in the second century. There is a record of fourteenth-century Emperor Hongwu ordering 15,000 sheets of perfumed toilet paper for his household. However, the first modern commercial loo roll was introduced by Joseph C. Gayetty of New York in 1857. He described it as 'the greatest blessing of our age'. For those people used to making do with old newsprint and corn husks, he was probably right. He called it 'Medicated paper for the water closet'.

The Romans used a sponge on a stick soaked in vinegar or wine to wipe their bottoms, while sailors in the Royal Navy used the frayed end of a rope kept damp in a bucket of seawater. Absorbing stuff.

Since 1975 it has been mandatory to put a bidet into any new bathroom built in Italy.

The Chinese invented the first toothbrush. They appeared in the fifteenth century and were made from wild-boar bristles. The first mass-produced toothbrush was invented by William Addis under the brand name Wisdom, which still exists today. He came up with the idea in 1780, while serving time in prison for inciting a riot. Americans, meanwhile, get through 1 billion toothbrushes in a year.

Myth: The appendix in human intestines has no purpose.

For a long time it was believed that the human appendix was a vestigial evolutionary remnant that once served a purpose but was now only of interest when it got infected and needed to be lopped off. But, more recently, the appendix has been found to contain lymphoid cells which help the body fight infection, so it may play a role in the human immune system, producing defences to fight serious infections. It also produces and stores bacteria that play a role in digesting food. If we get a gut infection such as dysentery, these bacteria can be flushed from our bodies. At this point our appendix releases its store of these important bacteria, which repopulate our digestive systems.

Is it true gold can be found in human blood?

Yes. Gold constitutes about 0.02% of our blood. So it's not really worth having it all drained and extracted. Blood contains other metals, including chromium, zinc, nickel, magnesium and, of course, iron.

Myth: Blood is red because it contains iron.

Haemoglobin carries oxygen around your body and gives blood its red colour. And, as we have just seen, it contains iron. Outside the body, we know that iron rusts easily, to a reddish tint (see page 54). But it's not the iron alone that makes the blood red, it is its interaction with oxygen. The more oxygen that is bound to it, the redder your blood is. So when it leaves your heart enriched with oxygen, it is very red. When it returns, and the oxygen is depleted, it is much darker. This has led to another myth – that blood deprived of oxygen is blue. It isn't, but when seen through human skin veins returning blood with less oxygen can appear blue because of how light travels through skin.

Do some animals have blue blood?

Most animals have red blood, just like humans. However, there are a few exceptions. Some types of octopus, squid and crustaceans have blue blood because it contains a high concentration of copper. When copper mixes with oxygen, it turns blood blue.

Myth: Royalty has blue blood.

No. Obviously. Well, not unless all those bonkers conspiracy theories about lizards are correct. Which they aren't. But if you are unfortunate enough to suffer from a rare condition known as methaemoglobinaemia, then your blood can turn blue, along with your skin, lips and nails.

Myth: Blood can only be red or blue.

No, just look at the skink, a type of lizard. One genus of skink, *Prasinohaema*, has green blood due to a build-up of biliverdin. Biliverdin is a by-product of the liver. Humans make it too, but we excrete it through our intestines (it contributes to the colour of poo). *Prasinohaema* do not excrete biliverdin, so it builds up in their body, turning their blood green.

The average human male has about 6 litres of blood, while a female has about 4.3. So comedian Tony Hancock was probably about right to complain that when he was donating a pint of blood, he was losing an entire armful.

Myth: Doctors stopped using leeches in the nineteenth century.

Well, yes, actually they did. But then we started using them again. The blood-sucking parasites, once used for blood-letting, which physicians believed helped to extract infection, were found to be no more than quackery. Unsurprising really, since they were removing vital blood from sick patients, who as a result just got, erm, sicker. However, over the last 30 years they have made a comeback in plastic and reconstructive surgery, and are used to restore blood flow to areas of damaged veins after skin or other tissue grafts, or where an appendage has been reattached. They can even be applied to the faces of people who have undergone cosmetic surgery. The European *Hirudo medicinalis* and Mediterranean *Hirudo verbana* species are the ones most frequently used in medicine. They have three saw-like jaws, with around 100 teeth each, which the leech uses to puncture the skin. It takes around two years from birth before a leech is ready for use on a human and they are starved for the last few months so that when they are put onto your skin, they are really, really hungry.

Leeches are the only animal domesticated for medicinal purposes.

But don't forget the maggots. They are not specifically domesticated but the rather unpleasant-looking larvae of the greenbottle fly are introduced into necrotic or gangrenous wounds, where they consume dead and infected tissue. The maggots release chemicals into the tissue that turn it into a liquid form they can digest, also consuming any bacteria present. They are tiny when first put onto the wound but grow up to about 12 millimetres in length as they munch away. Yum.

Why does there seem to be a delay in the pain arriving when I stub my toe?

This is due to the different kinds of nerve fibres involved. Sensory information travels fastest on the type that connects the touch receptors to your brain. The result is that you know you stubbed your toe a moment before the second type of fibre conveys the information to your brain that it's bloody painful. Ouch!

QUICK-FIRE FACTS

Around 3,000 different plant species are used in medicine.

Every two seconds somebody, somewhere in the world, needs a blood transfusion.

There are four main blood groups (A, B, AB and O), but all can be Rhesus positive or Rhesus negative, meaning there are eight in total.

Blood group O is the most common. Around 48% of the UK population has it.

Blood type O negative can usually be used in transfusions for almost everybody, because it has no antigens and so does not trigger an immune response from recipients.

There are other blood types outside the main ones. The rarest is RH null (or the golden blood type), possessed by fewer than 50 individuals in the world, most of them Australian Aboriginal people.

Australian James Harrison, who owed his life to a 13-litre blood transfusion at the age of 14, donated blood a record

HEALTH AND ILLNESS

1,173 times. His blood contained unusually strong and persistent antibodies against haemolytic disease found in newborn children. It is estimated he has saved more than 2 million lives.

The human brain grows three times its size during the first year of life.

There are more nerve cells in the human brain than there are stars in the Milky Way.

Cancer in the heart is rare because heart cells do not divide, a common prerequisite for the growth of cancer cells.

Most of our body waste – about 70% – is excreted through our lungs.

Arms are the most commonly broken bones, accounting for half of all breaks.

Researchers at the Massachusetts Institute of Technology measured cough droplets travelling up to 8 metres, although most are believed to travel only 2.

They also discovered droplets can stay in the air for up to ten minutes.

According to the UK's Alzheimer's Society, 1.6 million Britons will be living with the disease by 2040 (out of a population of about 70 million).

YAWNS FREEZE YOUR BRAIN

American football players are between three and four times more likely to suffer from Alzheimer's than the general population.

You will most likely die from sleep deprivation before you die from starvation.

Two out of three people do not have perfect vision.

The stuff you find in your belly button is dirt, grime and dead skin cells. Even worse, it provides a happy home for almost 70 species of bacteria. Remember the earwax and bogey rules (see page 41 and page 46). Don't eat!

One person in 100 has an extra rib, called a supernumerary, cervical or accessory rib. They occur most in males and quite often go undiagnosed, usually showing up in X-rays taken for other reasons.

Influenza causes more than 400,000 hospitalisations annually in the United States alone.

More than two-thirds of people diagnosed with cancer survive for more than five years, 50% for more than ten. However, survival rates vary from 98% for testicular cancer, to just 1% for pancreatic cancer.

Over the course of human history, only two diseases have been successfully eradicated: smallpox in 1980 and rinderpest in 2011.

HEALTH AND ILLNESS

When scientist Stephen Hawking was diagnosed with motor neurone disease in 1963 at the age of 21, he was given two years to live. He survived until he was 76.

Testicles are 2–3 degrees Celsius colder than the rest of the body, which aids sperm production. The average male makes about 200 million every day.

Women's ovaries, on the other hand, usually produce one egg per month.

Under the UK Public Health (Control of Disease) Act 1984, it is against the law for a taxi driver to convey a passenger who is suffering from the plague (they also can't take anybody with smallpox, typhus or cholera).

CHAPTER 8.

GEOGRAPHY

(or Why is the sea salty?)

Why is the sea salty?

Because the rivers that flow into it wash salts and other minerals out of the soil and rocks. These dissolve into the water and are carried into the sea. Other salts are added through hydrothermal vents on the seafloor. But why aren't rivers and lakes also salty? That's because the lakes and rivers are constantly emptying and refilling but the sea is a repository. The sea is also a far larger body of water, and the sun evaporates far more water from the sea than it does other water bodies, making the sea saltier than rivers or lakes. The most common salts in the sea are chloride and sodium.

GEOGRAPHY

Myth: The sea is blue because it reflects the colour of the sky.

It's not. Reflection accounts for only a tiny part of the sea's colour. It's mainly blue because water absorbs most of the light that strikes it, except for shorter wavelengths, which it scatters. Blue light has shorter wavelengths. Even distilled water is slightly blue because of this, but the impurities in seawater further enhance the blue scattering. Ice also retains this property, which is why icebergs have a blue tinge.

Myth: Mount Everest is the nearest point on Earth to space.

It isn't. That's because the Earth isn't a sphere. The force of its rotation – and because it's not solid on the inside – means that it bulges at the equator and is flattened at the poles (see page 30). The diameter at the poles is about 12,714 kilometres and at the equator it is about 12,756 kilometres. This means that although Mount Chimborazo in Ecuador is about 2,600 metres shorter than Everest from base to summit, its peak is nearer to the Kármán line (the boundary between Earth's atmosphere and outer space) because it is situated close to the Earth's bulging equator. By climbing Chimborazo you will be closer to space than any other human standing on our planet. But it's probably easier to book a flight over Ecuador than it is to climb Chimborazo.

It's estimated there are around three tonnes of human waste on Everest, most of it between Camp 1 at 6,065 metres and Camp 4 at 7,920 metres. The summit is at 8,849 metres. The problem became so bad that toilet tents were set up along the climbing routes but, when the mountain began to smell and climbers began to fall ill, in February 2024 the authorities decreed that all climbers had to bring their excrement back in bags, which would be checked at base camp.

There are 96 bags of human poo on the Moon left by the Apollo astronauts. Human biologists are keen to get their hands on it to see how much it has altered over the past five decades. Other things the astronauts left behind include a Bible, golf balls and photographs (probably now bleached white by the Sun's radiation).

Why are there mountains in some places on the Earth and not others?

The Earth's crust is made up of a series of 'continental plates' floating on a semi-molten mantle beneath. Where these plates collide, one can slide under another, but if they are of similar thickness and weight, neither will sink. Instead, they crumple and fold, and the rocks are forced upwards to form a mountain range. As the plates continue to collide, these mountains will get taller and taller. Formation of ranges such as the Himalayas and the Alps can take tens of millions of years.

GEOGRAPHY

What time is it at the North Pole?

Or the South Pole, for that matter? It's a good question because if you are standing with both feet on one of the poles, you are straddling all of the Earth's 24 time zones at once. What usually happens is that people working at the poles adopt the time zones of their departure nation. And, because at the poles there are three months of darkness in winter and three months of daylight in summer, they use their bodies' circadian rhythms to calculate how long a day is. It's also possible, although perhaps not so practical, to use the stars in the winter night sky to determine the time. As the Earth rotates these will appear to move, and when they return to their original positions, you know 24 hours have passed. You could do a similar thing with the Sun in the summer months.

The fact that at the North Pole all the time zones coalesce explains how Santa Claus manages to deliver presents throughout the world in just a single night.

Could a polar bear survive in the Antarctic?

Clever people know that polar bears are only found in the Arctic. But as their habitat shrinks, due to global warming melting the sea ice, would they survive relocation to the South Pole? They probably could, even though Antarctica is much colder. However, moving them there could have a drastic effect on the rest of Antarctica's wildlife. There are no predators like polar bears in the

Antarctic, which would mean that animals such as seals and penguins have not evolved to be scared of them. This could be devastating for their populations. As with any human intervention into the natural world, the possibility for unintended consequences is huge. Just ask Australians about rabbits.

Okay, ask me instead. When 13 European rabbits were released into the wild by Thomas Austin on his farm in Victoria in 1859, he had no idea that ten years later there would be hundreds of thousands. They might have been cute, but they devastated Australia's native vegetation to the point where the country's soil was being eroded. The Aussies built rabbit-proof fences (the longest was 3,256 kilometres) but it didn't stop them. By the 1940s there were 600 million in Australia and scientists had to introduce a virus, myxomatosis, to cull them. Lots died but they eventually became resistant to it. So another disease, calicivirus, was introduced in 1995. The rabbits have become immune to that too and are still eating their way across Australia. Think twice before moving that polar bear...

Which country has the highest life expectancy at birth?

According to figures published in 2024 by World Population Review, the country with the highest life expectancy at birth is Monaco, where you can expect to live until you are 87. Considering Hong Kong, Macau,

Liechtenstein, Vatican City and Singapore are also in the top ten, it seems that if you want your offspring to live a long life, move to a tiny country. The life expectancy at birth in the UK is 82.3, while in the US it's 79.7, in Germany it's 82.2 and in France it's 83.4. In Russia it's 74.6.

Why do people in Monaco live so long?

Health officials attribute the country's lifespans to three factors. The first is its Mediterranean diet, high in seafood, fruit and vegetables. Second is the country's state-funded healthcare system, and third is a very large percentage of wealthy residents attracted by tax breaks, meaning they can afford healthy lifestyles and any additional private healthcare needs they may require.

Which country has the lowest life expectancy at birth?

All the bottom ten are in Africa, with Chad and Nigeria filling the bottom two places, where you can expect to live until you are 54. However, there is some better news for Africa, where some countries have seen life expectancy increase by 10.3 years since 2000, against an average worldwide increase of 5.5 years.

What is the world's most common language?

It's Chinese (or a number of languages including Mandarin that share the same writing system) with 1,302 million speakers. The other nine languages in the top 10 are Spanish (427 million speakers), English (339), Arabic (267), Hindi (260), Portuguese (202), Bengali (189), Russian (171), Japanese (128) and Lahnda (117). German is 13th (77) and French 14th (76).

How did we end up with so many languages?

Around 7,000 different languages are spoken around the world, but nearly all belong to a small number of language families. It is believed that the languages in each family have a common ancestor. But when speakers of these ancestor languages began to move and live apart, they invented new words, phrases and eventually whole new languages. For example, even though French, Spanish and Italian are different languages today, they all come from a single language used a long time ago, known as Vulgar Latin. And you thought Vulgar Latin was a dictionary of swear words used in Italy, didn't you?

When did humans start talking?

Until very recently it was believed humans developed language about 200,000 years ago, but in March 2024 British archaeologist Steven Mithen of Reading University suggested that early humans in eastern or southern Africa known as *Homo erectus* invented a rudimentary language an astonishing 1.6 million years ago. Professor Mithen bases his estimate on the development of sophisticated stone tool technology. To transfer the complex knowledge and skills needed for these new tools from generation to generation would, he believes, require the existence of speech. He also argues that early humans would have needed language in order to leave Africa, a process that was happening at the same time.

Myth: The Welsh for a microwave oven is popty-ping.

It's a lovely story because *popty-ping* is beautifully onomatopoeic, but sadly it's apocryphal. It's true that *popty* is Welsh for oven, but the correct Welsh for microwave oven is *popty meicrodon* (*meicrodon* is Welsh for microwave and not a small Mafia boss, as some Welsh people jest). Walesonline explains that nobody in Wales uses *popty-ping* (the pinging oven) but people outside Wales like to think they do, which means it has taken on a life of its own. It's still a pity, though, that it's not true.

However, a ladybird in Welsh is a *buwch goch gota*, which translates into English as a short red cow, while a jellyfish is a *pysgodyn wibli wobli*, which translates as wibbly wobbly fish – that partially makes up for the disappointment of *popty-ping*.

Translation does, unsurprisingly, sometimes lead to unintended errors. When KFC translated their 'Finger Lickin' Good' slogan into Mandarin it initially said, 'Eat Your Fingers'. That's until 1.138 billion people pointed it out.

Other examples of woeful advertising translations:

When Coca-Cola wanted to appeal to the native population of New Zealand, they translated their slogan 'Hello, mate' into the Maori 'Kia ora, mate'. Unfortunately 'mate', in Maori, means death.

Ford found that their Pinto car had disappointing sales in Brazil. Turns out that 'Pinto' in Brazilian Portuguese slang means male genitalia.

Parker's leak-proof pen had the English slogan 'It won't leak in your pocket and embarrass you'. But the translator confused 'embarrass' with the Spanish word 'embarazar', meaning that the slogan now read: 'It won't leak in your pocket and impregnate you'.

When the company Powergen wanted to set up its Italian website, they gave it the domain name powergenitalia.com.

If you say it quickly enough ChatGTP sounds like 'cat fart' in French.

QUICK-FIRE FACTS

Concentrations of greenhouse gases, such as carbon dioxide and methane, are at their highest atmospheric levels for 2 million years. As a result, the Earth is about 1.1 degrees Celsius warmer than it was in the 1800s.

97% of climate scientists agree that humans are the cause of global warming and climate change.

NASA estimates that Greenland lost an average of 270 billion tonnes of ice per year between 1993 and 2019, while Antarctica lost about 145 billion tonnes of ice per year.

The ice sheet covering the continent of Antarctica contains 90% of the world's surface fresh water.

The global sea level rose about 20 centimetres in the twentieth century. The rate has doubled in the twenty-first century.

The highest ever tsunami to hit land was the one that struck Lituya Bay in Alaska in 1958. The average height of a tsunami once it hits land is around 30 metres, this one was 520 metres. Only two people died.

The deadliest modern tsunami was the one that struck countries surrounding the Indian Ocean in 2004. The death toll is estimated at 227,898.

Continents shift at about the same rate as fingernails grow.

There is enough gold in the Earth's core to cover the entire surface of the planet in a 50-centimetre-thick layer.

Since records began it has only snowed three times in the Sahara Desert, in 1979, 2016 and 2018.

California is the closest US state to Hawaii, but Hawaii is the furthest US state from California.

France's longest border, at 730 kilometres, is with Brazil. This is because French Guiana is fully integrated into the French Republic and is officially a part of France. It is also, therefore, a part of the European Union.

Other parts of the European Union not in Europe are Guadeloupe, Martinique, Mayotte, Réunion and Saint Martin (all French), the Canary Islands (Spanish) and the Azores and Madeira (Portuguese). As integral parts of the EU, they must apply its laws and obligations despite their diverse locations, which include South America, the Caribbean and the Indian Ocean.

More than half of the United States' coastline is in one state – Alaska.

Norway, the 61st largest country in the world, has a longer coastline (100,915 kilometres) than Russia, the largest country (37,653 kilometres), thanks to its hundreds of fjords and islands.

Norway is to the north, the south, the east and the west of its 'eastern' neighbour Finland (go check a map).

GEOGRAPHY

Russia (almost 17.1 million square kilometres) has more surface area than Pluto (16.7 million square kilometres).

Australia has 10,685 mainland beaches.

If deserts are defined as dry areas with little rainfall (and indeed they are), the largest desert in the world is Antarctica. Deserts needn't be hot.

South Africa has three capital cities. Cape Town is the legislative capital, Pretoria the administrative capital and Bloemfontein the judicial capital.

Death Valley is the driest place in the United States but in February 2024 a lake formed there, 86 metres below sea level, due to exceptional rainfall.

Africa is the only continent to straddle all four hemispheres.

Russia and China share the record for having borders with other countries – 14 each, including each other.

There is a town called Poo in Asturias, northern Spain.

Sudan has more pyramids than Egypt.

The largest pyramid in the world is the Great Pyramid of Cholula, also known as Tlachihualtepetl, in Mexico.

CHAPTER 9.

THE INTERNET AND SOCIAL MEDIA

(or Why is social media so addictive?)

Why is social media so addictive?

Social media activates the same reward pathways in the human brain that are triggered when we use any addictive substance. Drugs, including alcohol – however damaging they might otherwise be – release the feel-good chemical messenger dopamine, as does social media. Dopamine acts on the brain to give feelings of pleasure, satisfaction and motivation, and also helps control memory, mood, sleep and concentration. Spend more and more time on social media platforms and the desire to repeat the dopamine hits increases. Put down that phone...

It's estimated that as many as 10% of Americans are addicted to social media. Similar to drug abuse, social media addicts go through withdrawal when the source of their addiction is removed. Symptoms of withdrawal include anxiety, depression, mood swings, indifference to the real environment, fear (mainly of missing out on something important) and insomnia.

But why is social media so popular in the first place?

Well, indeed. We interacted with our fellow humans before it existed, so what has changed? Apparently, it's the way the interactions between users are presented. David Greenfield, a psychologist who studies technology addiction, told *The New York Times* that users are searching for a reward that might come at any time, so it is unpredictable, just like success when playing slot machines, however it is even more addictive because 'users are beckoned with similar lights and sounds but also powerful information and reward tailored to the user's interests and tastes'. Young people are particularly susceptible because the adolescent brain is especially attuned to making new social connections.

There are almost 4 billion active social media users (out of a world population of 8 billion). A new social media account is created every 6.5 seconds.

In 1990, Mike Godwin, an author and attorney, came up with 'Godwin's Law of Nazi Analogies', which states: 'As an online discussion grows longer, the probability of a comparison involving Nazis or Hitler approaches 1.'

Myth: Tim Berners-Lee invented the internet.

In 2004 Briton Tim Berners-Lee received a knighthood and henceforth became known as 'the man who invented the internet'. And it would be most unfair to say that his work wasn't very, very important in the creation of the ubiquitous tool pretty much all of us use in some form every day. But long before Berners-Lee's ground-breaking work, an engineer at the University of California, Leonard Kleinrock had been asked by a company called the Advanced Research Projects Agency (ARPA) to devise a system that would link together in a network all of its researchers' computers. That was in 1966, and in 1969 this happened:

What was the first message sent over the internet?

It depends how you define the internet (see above and below) but on 29 October 1969 the first electronic message was sent between two computers on the network Kleinrock had devised and which he had called ARPANET. The programmers had intended to type and transmit the word 'login', but the system crashed after the first two letters. So the first message sent was actually 'lo'. An hour later the other three letters arrived.

The first email (defined by the first use of the @ symbol) was sent two years later, in 1971, also using ARPANET, by Ray Tomlinson of the company Bolt, Beranek and Newman. Now 262 billion emails are sent every day.

THE INTERNET AND SOCIAL MEDIA

In 1998 journalist Paul Krugman wrote that 'by 2005 or so, it will become clear that the internet's impact on the economy has been no greater than the fax machine's'. Ooops.

But have you finished the Tim Berners-Lee story yet?

No. ARPANET was a clever idea but it kept crashing. So in 1975 Bob Kahn and Vint Cerf at Stanford University came up with what is called TCP/IP (Transmission Control Protocol/Internet Protocol) which tidied up the transmission errors that ARPANET was causing. When the US Department of Defense and telecoms company AT&T began using TCP/IP and offered it free to the public, everybody else began to adopt it. Which is where Sir Tim comes in. He was working at the particle physics laboratory at CERN in Switzerland and was frustrated that, although computers could now be successfully linked together, he couldn't easily view, share or send links to documents. In 1989 he proposed new software called the WorldWideWeb. This described how a 'web' of 'hypertext documents' on 'web servers' that could be viewed by 'browsers' might work. In 1990 the world's first website was up and running at info.cern.ch (it contained the laboratory's telephone directory). So although Sir Tim didn't invent the internet, he did invent the WorldWideWeb, which ensured ordinary people – like me and you – with absolutely no knowledge of computer systems, could use it.

Why are attention spans becoming shorter?

They probably ar... er... What?

They probably aren't, but the way we consume information is. In the age of the internet we are constantly bombarded with it from so many sources. This has led some researchers to conclude that we only have time to grasp the gist before we move onto something else. We need to know in moments whether the information will be worthwhile or not (just look at all those people swiping left at incredible speed).

Until the dawn of the internet our rate of information uptake was, by default, much slower. There was time to pause and ponder. Apparently, the average person living around 150 years ago experienced as much information in a lifetime as the average person now does in a year. Similarly, a person living in the Middle Ages consumed as much information in a lifetime as we now do in a day.

Why do people fall for internet scams?

The University of New South Wales has identified six key reasons why people fall for scams on the internet. The top one is financial desperation, followed by naivety and a lack of awareness. Third is social engineering – scammers manipulate their victims and create a false sense of trust. Fourth is people believing that the scams are originating from people in authority, such as the government or a trusted bank. Fifth is a lack of vigilance – in a rush we overlook warning signs or suspicious behaviour. And lastly there are emotional triggers – sadly, too many of us simply want to fall in love, and it clouds our judgement.

The median loss across all romance scams in the US in 2021 was US$2,400 (£1,930) – but this amount was more than three times higher among adults aged 70 and over.

What are the worst internet scams?

Okay, so it might not be the worst, but it's probably the most enduring. You get a message from a Nigerian prince who needs help reclaiming a vast sum of money he has been cheated out of. If you help him by sending him money (and not just once) to aid his cause, he'll reward you a million times over. The fraud has been running for years and has swindled tens of thousands of people out of their fortunes, and continues to do so. The scam has even been updated. There's a Nigerian astronaut on an abandoned Russian space station who really needs your help getting home...

Here are a couple more:

It's not just old folks who are preyed on by internet scammers. Many millennials (a person born between the early 1980s and the late 1990s) live their lives pretty much online. And some can be quite vain, it seems. A scam to steal their online details told them that there was a 'Hot List', based on their physical appearance. And if they handed over their Instagram account details they would see their name on it. They did... and they didn't.

In the early years of the internet, many people were sent a website popup message which told them they'd won a cash prize for being the 100,000th visitor to the site and requesting a small cash token to process the prize money. Guess what happened next? This scam made a comeback in 2023, presumably on the basis that the scammers thought we'd all forgotten about it. And some people had.

Elon Musk is even in on it (except he isn't). But you might get an email from elonmuskdonation99@gmail.com offering you 10 Bitcoin worth 4,124,270.00 (although it doesn't actually say 4,124,270.00 what). Again, you have to send 'Elon' a 'release fee' and he'll send you the Bitcoin. In 2022 more than 12,000 people did just that.

Just as did many parents who, in 2022 and 2023, received a text from their offspring saying: 'Mum, my phone has broken and this is my new number x.' They then asked said parent to send money direct to the account of their landlord because they'd lost all their bank details, which had been on their old, broken phone and now they couldn't pay their rent. Apparently 11,000 Australians fell for it and handed over a total of 7.2 million Australian dollars.

Meanwhile, Zeltenis says he is an offshore oil-rig operative. When 'he' meets people online he falls deeply in love with them and asks for £3000 so he can leave the rig and visit them. They send it and – surprise, surprise – Zeltenis never turns up. However, Becky Holmes wasn't prepared to wait for Zeltenis. She told him there was no need for her to send £3000 because she was coming to find him. Becky was also an undercover journalist who exposes fraudsters. She took a boat and then a helicopter, constantly updating Zeltenis on her journey and how close she was to him. His last message was clearly written in a panic: 'Coming to sea is last thing I allow you to do so long you are my woman. Calm your head and send me fee.' Fortunately, some people fight back. We admire Becky.

And yet people still fall for scams. We say never, ever pay anything to anybody you have never met in person. And even then exercise extreme caution...

More than seven in ten internet users have been targeted by scammers.

How much does the internet weigh?

Apparently, about 60 grams. That's the weight of the electrons active at any one time that are necessary to sustain the global network. Considering it takes about 2 billion electrons to produce a single email, that's a lot of electrons.

When was the first mobile phone call made?

The first ever mobile phone call was made on 3 April 1973, by Motorola engineer Martin Cooper, while standing on the corner of Sixth Avenue in New York City. He called a rival at Bell Laboratories to announce he was speaking from 'a personal, handheld, portable cellphone'.

The very first YouTube video to be uploaded was 'Me at the zoo' on either 23 or 24 April 2005. It's 19 seconds long and features YouTube's co-founder Jawed Karim in front of two elephants at the San Diego Zoo in California.

What's the internet's most searched-for medical condition?

According to Google, in 2023 the most searched-for health condition worldwide was diabetes, closely followed by cancer. Neither were more specific, and in third place was the term 'pain', which, obviously, can cover any number of conditions. At the very bottom of the list was lichen simplex chronicus, which (if you, errr, google it) apparently is a chronic itchy skin disorder.

How many people misdiagnose themselves on the internet?

It's called cyberchondria – searching for what's wrong with you on the internet and coming up with entirely the wrong conclusion. And a lot of us do it. According to health company LetsGetChecked, 65% of Americans have attempted to diagnose themselves on the internet and of those 40% come to the wrong conclusion. Even worse, 75% of them said searching for a diagnosis made them even more stressed than they had been. Etactics.com says that online symptom checkers only give a correct diagnosis 34% of the time, while real-life doctors have an accuracy rate of almost 90%.

67% of people who use online symptom checkers are women, while adults aged between 18 and 29 are the most avid users of health websites.

Self-diagnosis has led to a 5% drop in GP visits in the US. So, although you shouldn't do it, it appears to save money.

QUICK-FIRE FACTS

The most popular social media platform in the world is Facebook, with 3,049 million active users in 2023.

75% of the world's women use Facebook (63% of men do too).

Only 25% of Facebook users check or adjust their privacy settings.

The most followed person on Facebook is Portuguese footballer Cristiano Ronaldo.

YAWNS FREEZE YOUR BRAIN

The other social media platforms in the top four, behind Facebook, are YouTube, WhatsApp and Instagram.

The average user spends 142 minutes on social media every day (or 15% of their time awake).

The average American checks their phone 46 times per day.

The average person watches 19 minutes of YouTube videos a day.

A billion hours of videos are watched on YouTube every day.

More than 70% of what users watch on YouTube is determined by the site's recommendation algorithm.

350,000 of what were once called tweets (Xs?) are sent every minute.

Larry Bird is dead. (He was the name for the old Twitter logo before it was replaced by an X.)

In January 2024, the world's most popular internet search term on Google was YouTube. Others in the top 100 included Weather (4th), Satta King (a form of illegal gambling prevalent in India) (15th), Cricket (41st), Calculator (56th) and Daily Mail (67th).

THE INTERNET AND SOCIAL MEDIA

The first prototype computer mouse was built in 1964 by Douglas Engelbart of the Stanford Research Institute. It was first demonstrated to the public on 9 December 1968.

The plural of computer mouse was originally computer mouses. Many people now call them mice.

A computer mouse is called a mouse because originally it was linked to the computer via a wire which resembled a tail, while the on-screen cursor it controlled was once known as a CAT. Wireless mouses (mice) have reduced the resemblance to rodents.

The first multiplayer game on the internet was Multi-User Dungeon (MUD), introduced in 1978.

The first online edition of a newspaper was the *Columbus Dispatch (Ohio)* on 1 July 1980.

The first computer virus was the Creeper, which infected machines in 1971 with the message 'I'm the creeper: catch me if you can'.

Although Kevin Mackenzie is credited with creating the first emoji -) in 1979, it wasn't until Scott Fahlman created :-) that their use took off.

The first spam email (actually sent on ARPANET – see page 104) came from computer salesman Gary Thuerk on 3 May 1978. He was advertising his forthcoming presentation on Decsystem-20 products. When it was sent, the terms email and spam didn't exist.

Malware and internet bots account for two-thirds of internet activity.

The real name for a hashtag is an octothorp.

More people own a mobile phone than own a toothbrush.

CHAPTER 10.

HISTORY AND POLITICS

(or Why do we have leap years?)

Why do we have leap years?

The Earth doesn't take 365 days to orbit the Sun, it takes 365.24219 days (or almost 365 and one-quarter days). This means to make up for those quarter days we add one day (29 February) to the calendar every four years. However, because they are not quite full quarter days, every so often we have to skip what should be a leap year. So if the year is divisible by 100 but *not* divisible by 400, we skip the leap year. This means that although 1700, 1800 and 1900 weren't leap years, 2000 was. And 2100 won't be. If we didn't have leap years at all, in about seven centuries our summers would start in December and we'd be eating Christmas turkey in June.

The Romans had figured this out but hadn't realised that a year was just a tiny bit short of 365 and one-quarter days (44 minutes in fact). So, over time, their calendar began to drift and the equinoxes, solstices and important festivals were starting to fall at the wrong time of year. In October 1852, ten days were dropped from the calendar at the decree of Pope Gregory XIII to make sure everything was lined up again, which is why we now observe the Gregorian calendar.

Except, except, except... the Orthodox Church didn't want to. It stuck with the old Roman Julian calendar proposed by Emperor Julius Caesar. And that's why Russian Orthodox Christmas Day is on 7 January, not 25 December. And smart readers will have noticed that 7 January is 13 days after 25 December, not 10. That's because since October 1852 the Gregorian and Julian calendars have drifted even further apart. Before 1923, Russian Orthodox Christians celebrated on 6 January and – you guessed it – after 2100 they'll celebrate on the 8th.

People born on 29 February are known as leaplings. Famous leaplings include composer Gioacchino Rossini, former Indian prime minister Morarji Desai, actor Joss Ackland, rapper Ja Rule and Pope Paul III (who probably cursed his predecessor Gregory). Leaplings have their own special cocktail with which to celebrate the day. For the leap year of 1928, Harry Craddock, bartender at London's Savoy Hotel, invented a cocktail especially for them. Mix together gin, Grand Marnier, sweet vermouth and lemon juice. Don't miss out because it'll be another four years before you can have another.

HISTORY AND POLITICS

Who was the first female leader of a nation?

Excluding monarchs and including only elected individuals, the first woman to become head of state anywhere in the world was Khertek Anchimaa-Toka of the now defunct Tuvan People's Republic in 1940. The Tuvan PR was a state that existed between 1921 and 1944 and is now an administrative part of Russia. However, while Anchimaa-Toka is generally accepted as the 'first ever elected female head of state', the election which put her there was not considered free or fair (and she was the wife of the nation's supreme leader).

So who *was* the first female leader of a nation?

If we are discounting Khertek Anchimaa-Toka, who seems to have cheated a bit, then it is Sirimavo Bandaranaike of Ceylon (now Sri Lanka), who was elected prime minister in July 1960.

The first woman to become president of a country was Isabel Perón, who became president of Argentina in July 1974. However, she was not elected. She had been serving as vice-president to her husband and succeeded to the president on his death. Therefore, the first elected female president of a country was Vigdís Finnbogadóttir of Iceland, who won that nation's 1980 presidential election.

'It will be years – not in my time – before a woman will become prime minister.' So said British politician Margaret Thatcher in 1974, five years before she became prime minister. Britain has now had three female prime ministers, with Theresa May and Liz Truss following in Thatcher's footsteps.

The nation that has had the most female heads of state is Switzerland, with five. Finland has had four. The longest-serving female head of government without leaving office was Angela Merkel, who led Germany for 16 years and 16 days. However, Sheikh Hasina, the former prime minister of Bangladesh, served for more than 19 years, although that was over two non-consecutive periods.

Why do we have different political beliefs?

Surely there's a right way and a wrong way to do things, experience over many centuries should have taught us this. Yet we still continue to argue, and elect politicians who best suit our views. Psychologists think that our personalities influence our politics. They watched how people behaved in their homes and work places before asking their subjects what their political preferences were. They found that the people who were conservative had tidier home and work spaces and owned more objects related to order, such as diaries or calendars or other organisers. Progressives or liberals had more cluttered spaces and objects related to learning about new ideas. Conservatives tended to be more punctual than progressives. The two groups also differed in their social preferences. Conservatives are more likely than progressives to admire people of high social status, or traditionally dominant societal groups such as White males and heterosexuals. Progressives, while still succumbing slightly to an admiration of high status, are less likely to care about ethnicity or sexuality. Conservatives dislike ambiguity, progressives find it interesting. Moral choices vary too. Progressives get angry at inequality and injustice, while conservatives are more bothered about disrespect for authority and tradition. They see the world as dangerous and have stronger reflexes to stimuli, such as loud noises or pictures of people dressed differently from how they dress. And just in case this is all sounding a bit intellectual, it's worth noting that conservatives are more disgusted by the smell of farts than progressives.

Geneticists think there may be a genetic basis to political belief (and not just to keep themselves in a job). They note that people with the 7R variant of the DRD4 dopamine receptor gene are more likely to be open-minded and have liberal political leanings.

The reason we call people left-wing or right-wing comes down to the French Revolution. In the legislative assembly in Paris the traditionalists supporting the king and church sat on the right side of the assembly while the revolutionaries sat on the left.

Myth: British elections are a secret ballot.

No, they are not and nor are those of many democracies. In the UK votes can be traced by matching the numbered ballot paper to its similarly numbered counterfoil. This counterfoil bears the voter's registration number, written there by the polling clerk who issues the ballot papers when you turn up at the polling station. Officially all the ballot papers, after being counted, are placed in sealed containers and destroyed after one year and one day, meaning that nobody can check who you voted for. But if somebody wanted to, in theory they could.

Myth: Diane Abbott, Paul Boateng, Bernie Grant and Keith Vaz were the UK's first ethnic-minority MPs.

All four entered Parliament after the 1987 general election. However, although records were not kept and the language of the times was subjective, recent research has suggested the first MP not from a white British background was James Townsend, who was elected MP for West Looe in 1782. It is believed he was of partial Black African ancestry. His maternal grandfather had married the daughter of a Black African woman and a White European soldier. Townsend had already served as the lord mayor of London.

Things we shouldn't have forgotten about history.

Anne Boleyn was Elizabeth I's mum.

Ada Lovelace – the world's first computer programmer – was the daughter of poet Lord Byron.

HISTORY AND POLITICS

American Harriet Quimby was the first woman to fly across the English Channel on 16 April 1912, but because the *Titanic* sank on the previous day, she has been pretty much forgotten.

In 1955, Claudette Colvin was arrested for refusing to give up her bus seat to a White person in Montgomery, Alabama nine months before Rosa Parks became a cause célèbre for doing the same thing.

Thomas Edison is regarded as the inventor of the light bulb, but Alessandro Volta, Humphry Davy, James Bowman Lindsay, Warren De la Rue, William Staite and Joseph Swan all invented versions before him.

In 1858, Alfred Russel Wallace wrote a paper explaining the theory of evolution before Charles Darwin released his. But Darwin had friends in high places who – seemingly against Darwin's wishes, and while Wallace was in the Malay Archipelago – insisted Darwin publish *On the Origin of Species*. He did so in 1859 and the credit for one of science's greatest theories now overwhelmingly lies with him.

We are still living in an ice age. We may think of ice ages as things of the past, with woolly mammoths marauding the frozen landscape, but we are living in one now. Despite the rapidly melting ice sheets in Greenland and Antarctica, we are in the Quaternary Ice Age, which started 2.6 million years ago. Currently the Earth is in what is known as an interglacial period, which began about 11,000 years ago and partly explains why it doesn't seem so ice-age-like.

Myth: All the historical events below actually happened.

Except they didn't, it's just that lots of people think they did.

An apple fell on Isaac Newton's head. It didn't. He probably did see one falling, however, which led to him deducing that a force – which he named gravity – was acting on it.

Revolutionary Protestant reformer Martin Luther nailed his list of grievances about the Catholic Church to the door of the church in Wittenberg in 1517. He didn't. What he did do, however, was write them down and then post the letter to the Archbishop of Mainz.

Nero fiddled while Rome burned. No, he didn't, he was in Antium 45 kilometres away.

Rats spread the Black Death. No, they didn't, and researchers at the University of Oslo have recently concluded that it wasn't even the fleas that lived on the rats that caused the plague. Fleas were, however, responsible. But it was the ones living on humans that caused the pandemic. Humans come into far more contact with other humans than rats ever do (or did).

Rosa Parks (see page 121) was sitting in a White section of the bus in Montgomery, Alabama when she was asked to move to let a White man sit down. No, she wasn't. She was actually in the African-American section of the bus,

HISTORY AND POLITICS

In 1850, US president Zachary Taylor died after eating too many cherries and drinking too much milk at a Fourth of July celebration. Doctors believe the acid in the cherries reacted badly with the milk.

Disgraced United States president Richard Nixon could play the piano, saxophone, clarinet, accordion and violin.

The first British prime minister to die in office was Spencer Compton, who succumbed to 'ill health and overwork' in 1743. Compton was the first Earl of Wilmington (in Kent) and the US city of Wilmington, Delaware is named after him.

Seven British prime ministers have died in office, but only one, Spencer Perceval, has been assassinated. In 1812, he was shot in the lobby of the House of Commons by John Bellingham, a Liverpool merchant who blamed the government for his personal financial losses.

Liz Truss is the shortest-serving British prime minister in history, racking up a mere 49 days in 2022.

Iceland has the oldest parliament in the world. The Althing was established in 930.

Pedro Lascuráin was president of Mexico for less than an hour in 1913. The previous president Francisco Madero had been tortured into resigning and the constitution deemed Lascuráin, the foreign secretary, become president. He really didn't want the job and jacked it in after 45 minutes. Excluding monarchies, it was probably the shortest term as leader anywhere in the world.

The shortest-serving monarch in history is probably Louis XIX of France. After his father's abdication during the July Revolution, on 2 August 1830 he became king, but abdicated 20 minutes later.

The heaviest monarch is believed to have been Tāufaʻāhau Tupou IV, King of Tonga from 1965 to 2006, who weighed 208.7 kilograms in 1976.

Marie Antoinette's son Louis was imprisoned after the French Revolution. He died aged ten but in the years that followed more than 100 people claimed to be him.

There were female gladiators in ancient Rome.

Pope Gregory IX declared in 1233 that domestic cats were associated with the devil and in a Papal Bull demanded that they should be exterminated.

All cats were included in his decree, but he did reserve particular ire for black ones.

Before he became Pope Pius II in 1458, Aeneas Sylvius Piccolomini wrote an erotic novel called *The Tale of Two Lovers*.

HISTORY AND POLITICS

In 1644, the Lord Protector of England and staunch Protestant Oliver Cromwell banned people from eating mince pies. He decreed they were a pagan form of pleasure. The ban lasted until 1660.

Former North Korean dictator Kim Jong Il composed six operas while he was in office.

Romantic poet Lord Byron kept a bear in his rooms at Trinity College in Cambridge University in the early 1800s. He did it because the college had rules forbidding dogs but not bears.

Queen Victoria could speak Hindi.

Adolf Hitler, Joseph Stalin and Benito Mussolini were all nominated for the Nobel Peace Prize.

In 1908 the state of New York banned woman from smoking in public. The law lasted only two weeks before being repealed, and only one woman was prosecuted.

CHAPTER 11.

SOCIETY AND RELIGION

(or Why do we make friends?)

Why do we make friends?

The average human has about five intimate friends, ten close friends, 50 true friends, 100 peripheral friends and about 1,500 people who are neither strangers nor close acquaintances. But why do we have them at all? Well, it would seem they are a biological necessity. Anthropologists think it's because we live in large social groups and, as these can create tensions between individuals, we need our own personal protection squads helping us to create alliances that maintain group stability. We also know that social isolation leads to stress and anxiety. People with good friendship networks are healthier physically and mentally, while our brain's reward centres produce oxytocin (known as the 'cuddle' or 'bonding' hormone) that is released during positive contact with other people, as are the neurotransmitters endorphins which offer a sense of wellbeing.

Myth: We choose our friends randomly.

Not so. Although we may choose friends who have similar interests, beliefs and personalities to us, an even stronger draw is genetics. Amazingly, most of our close friends are genetically related to us, quite often as close as a fourth cousin (that is, you share the same great-great-great-grandparent). Nobody knows quite why this is or how it happens but similarities in appearance, mannerisms and smell (and, obviously, geography) probably play a role.

Although we choose our friends, we also lose them, and at a faster rate than we do with family ties. If we don't see a friend for 12 months, the hierarchy of their friendship falls by about a third. While family connections remain pretty much throughout a lifetime, we tend to lose (and gain) 20% of our friends every few years.

Why do we hate?

It's complex but psychologists have a number of theories. We are wired to fear things and people who are different from us and, when we feel this perceived threat, we turn inwards to those people with whom we identify. This can lead to aggression towards the outsiders who seem different, dangerous and threatening, even though they may be benign. Other researchers believe that hate fills a void, helping to foster a sense of connection and camaraderie with others who dislike the same things. It distracts from the challenging requirement of creating one's own identity, especially if an individual feels helpless or powerless, or where there is a sense of injustice, inadequacy or shame. It is easier to project this on to blaming somebody else when you are (or feel) incapable of fixing the problems yourself. This explains why immigrants or outsiders are so often blamed for a nation's perceived ills.

But does hate have an evolutionary benefit?

It can have. If you realise that an individual or group of individuals will be detrimental to your status or health and may cause you injury, you can grow to hate them. And you may then opt to avoid them. This means you are more likely to pass your genes on than if you confronted them and were physically damaged by them. Distress is rarely rewarding for a human and so avoiding distress by discovering what you hate means you will avoid such people and situations to the benefit of yourself and possibly wider society. The author would like to make it clear that he usually avoids watching Manchester United playing football. It distresses him.

Why do we fall in love?

Well, indeed? Why do we? We know a bit about the chemistry of how it happens. It's down to three chemicals. The first, noradrenaline, stimulates the production of adrenaline causing our hearts to race. The second, norepinephrine, also known as the fight-or-flight hormone, creates butterflies in our stomachs. And the third, dopamine, also known as the feel-good neurotransmitter, gives us that oh-so-pleasurable feeling. It also makes us behave in illogical ways – so, yes, when you fall in love, you really are behaving irrationally. But what's the biological benefit of love? We know it's not required to make babies. So is there an evolutionary reason for love? Well eventually those early three chemicals are replaced by another: oxytocin (see page 128). This chemical ensures partnerships – fingers crossed – endure. And children brought up in enduring partnerships generally have better life outcomes. What's more – and this works for heterosexual and homosexual relationships – those in them have better health outcomes, fewer strokes, fewer heart attacks, less depression and better survival rates from cancer. And couples in long-term relationships are more empathic, compassionate and tender, and generally contribute more to society. Love conquers all, it seems.

In the UK, 42% of marriages end in divorce. In the US, it's around 49%, with a marriage ending, on average, every 36 seconds.

The most common Valentine's Day gifts in the UK are, in order, chocolates, flowers, fragrance and lingerie. Lack of imagination clearly runs them close, though.

St Valentine was a third-century Roman priest, who performed secret Christian weddings against the wishes of the authorities. He was martyred for his actions, supposedly on 14 February. As well as lovers, he is also the patron saint of epilepsy, beekeepers and the Italian city of Terni.

Which country is the happiest?

According to Gallup's 2024 World Happiness Report, which correlates the various factors affecting people's lives through interviews, it's Finland, closely followed by Denmark and Iceland. And the unhappiest? Afghanistan.

How much money do we need to be happy?

The average amount of income per year required for the vast majority of people receiving it to say they are happy is around £75,000 (€87,000 or US$95,000), which is around double an individual's average annual income in the UK. However, research by the United States Academy of Sciences has shown that if you earn more than this, you still get extra joy, but with progressively diminishing returns. If you earn ten times the happiness average (£750,000), you will be happier, but only by about 10% despite earning £675,000 more. For research purposes, the author is happy to be provided with such a sum just to find out if this is true.

Myth: Money can't buy you happiness.

Unfortunately, it seems that it can. The research mentioned in the previous answer has also shown that there is a clear link between having money and being happy. Only around 7% of people surveyed said that money made no difference to their level of happiness. And none of these people earned less than US$33,000 (or £26,000). Strangely enough, it seems that if you live in Manhattan or Mayfair, you are generally happier than if you live in Detroit or Doncaster (with all due respect to the fine cities of Detroit and Doncaster). Who would have thought?

Burundi is the poorest country in the world with a gross national income per person in 2022 of US$240 (or £190). Meanwhile, Norway is the richest country in the world with a gross national income per person of US$95,510 (or £75,750).

Why do women on average live longer than men?

In 2023, the United Nations said worldwide life expectancy for women was 76.0 years, but for men it was only 70.8. Women outlive men in nearly every society, with the biggest gap – 13 years – found in Russia. But why? Part of the answer is behaviour. Men are more likely to smoke than women, more likely to take risks, and more likely, even today, to work in jobs considered dangerous, making them more susceptible to injury. But biological differences have an effect too. Biologists believe the female hormone oestrogen combats numerous conditions, including heart disease. Women are also thought to have stronger immune systems than men. As becomes obvious when men contract flu – it's always worse.

Myth: The number of people who believe in a god outweighs the number of atheists 93% to 7%. Which means more people believe in a god than don't.

But is this true? Nearly all people who believe in a god only believe in their own specific god or deities. They don't believe in all of the others. A Christian does not believe in the God of Islam, nor the numerous Hindu deities and possibly only partially in the Jewish God. Christians would also think it ridiculous to believe that gods exist in volcanoes or in inanimate objects such as televisions, for example. Very few people believe in all the gods that have ever been worshipped and all those yet to be proposed. Calculated this way, there is far more disbelief in the world than there is belief.

People who believe in all gods and deities, including those yet to be proposed, are called omnists. Their belief is known as omnism. They are very few in number (the believers, not the gods), possibly fewer than there are described gods. True omnists make up about 0.000000001% of the world's population. That's not many people, probably because attending all those religious services is quite tiring. Notable omnists are basketball player Shaquille O'Neal and deceased jazz saxophonist John Coltrane.

The world's oldest temple is believed to be Göbekli Tepe in Turkey, erected about 11,600 years ago. It is the first evidence of organised religion, although what that religion was, remains unclear.

Why are there different religions when only one can be right?

Religions help some people answer very big questions, such as why we are here and who, if anything, created us. But why so many? People have different religions for the same reasons people have different opinions and different tastes, because they were raised in different ways, in different places, in different families and at different times. And we all think differently. All these things have an impact on what we believe to be true about the world. For many people, religion isn't so much about deities as it is about a sense of community and rules for how we interact with others. Research has shown that for many people religion is not about their specific god but instead is used as a means of understanding the world around them. So the actual religion matters less than the geographical area you were born into and how the majority of people in that geographical area organise their societies.

What's the strangest creation story?

It's pretty much take your pick. The Vikings celebrated the primordial giant Ymir. After he was formed from the icy rivers of Élivágar, he was fed for three days by the celestial cow Auðumbla, whereupon he created the first man and woman who dropped from his armpits. After he was killed, his blood became the Earth's seas, his hair the trees, his bones the mountains and his brains the clouds.

Oddly, his eyebrows became Midgard, where humans have lived ever since.

Meanwhile the Boshongo people, who live around the Congo River in central Africa, believe that their god Bumba had a terrible stomachache, which was only relieved when he vomited out the Sun. This evaporated some of the water in Bumba's otherwise sodden and dark world and produced the land. Bumba subsequently puked out the Moon, stars, numerous creatures and humans.

Scientologists believe that a human is an immortal, spiritual being (called a thetan) that is resident in our physical body. Thetans have had innumerable reincarnated lives, some of which, before they arrived on Earth, were lived in extraterrestrial cultures. By undergoing auditing (a series of classes), Scientologists believe that people can free themselves of their human form and reclaim their true selves. Once you have finished auditing you can become closer to the Supreme Being and unlock special powers usually repressed by being an everyday human. However, the founder of Scientology, science fiction (sic) writer L. Ron Hubbard, also once said, 'If a man really wants to make a million dollars, the best way would be to start his own religion.' Those classes cost money. So it might pay to keep an open mind about thetans.

Meanwhile Jainism, practised in India, is far more pragmatic. It does not believe in a creator. According to Jain doctrine, the universe and its constituents have always existed and everything is governed by universal natural laws. Jain philosophers believe that a conscious, immaterial entity such as a god could not possibly make a material entity such as the universe.

Myth: You need to see a priest if you want to confess.

It probably depends how devout you are, but not necessarily so. In 1993, Greg Garvey invented an automatic confession machine. You entered the date of your last confession, the number of sins you'd committed (with details) and the machine calculated your penance and delivered a printout of what steps you should subsequently take to achieve absolution. Priests, worried about imminent unemployment, pointed out the machine had not been ordained so it didn't count.

QUICK-FIRE FACTS

The first city is believed to have been the Sumerian city of Uruk in Mesopotamia. It dates back more than 6,000 years. It housed about 50,000 people.

The number of people living in cities all over the world overtook the number living in rural areas in 2014. Just four decades earlier, only a third of people lived in cities.

Cities occupy less than 1% of the Earth's surface. They could all fit onto the island of Borneo (about 750,000 square kilometres).

The oldest intact shoe dates back 5,500 years. It was found in a cave in Armenia. But shoes predate that – foot bones and cave drawings show that 40,000 years ago people were wearing shoes in China.

The first evidence of humans wearing clothes is 170,000 years ago, long before humans emigrated out of Africa.

Ruben Enaje, a carpenter from the Philippines, has been nailed to a cross at Easter 35 times. He is the world's most crucified person but said in 2024 that perhaps he was getting a bit too old for it.

According to a survey by YouGov, 2% of Church of England priests don't believe in God.

Another 16% are agnostic.

Although it is very difficult to pinpoint, it is believed that Hinduism is the world's oldest mainstream religion. It has texts dating back 3,000 years.

However, Zoroastrianism, the ancient religion of Persia, which is still practised today, also has roots in Hinduism, so can claim to be equally old.

And Judaism, although with no formal written records for proof, has an oral tradition that may even predate Hinduism and Zoroastrianism.

Months that begin on Sundays will always have a Friday the 13th.

In 1973, the first female stockbrokers were allowed onto the floor of the London Stock Exchange, 275 years after it first opened.

There are pig cuddling cafes in Japan. Customers pay up to US$15 to cuddle miniature pigs for 30 minutes.

A word that is only used or found once in the written record of an entire language is called a *Hapax legomenon*. It means 'said only once' in Greek.

Examples in English include 'hebenon', the name of a poison in William Shakespeare's *Hamlet*, 'sassigassity', meaning audacity and used by Charles Dickens in his short story *A Christmas Tree*, and 'nortelrye', a word used by Geoffrey Chaucer to mean education.

There are many *Hapax legomena* (yes, that's the plural) in the Bible. St Paul's Epistle to the Romans has 113, and his First Epistle to the Corinthians has 110. Whether the Romans or Corinthians understood any of these words they'd never encountered before remains unknown.

Japanese, too, has many characters that have only ever been used once in the entire written language. They are called *kogo*, or 'lonely characters'. Because some were used a very long time ago, scholars aren't even sure what they mean.

Of the languages that use letters, the one with the longest alphabet is Khmer (or Cambodian), with 74 letters.

Rotokas, a language spoken in Papua New Guinea, only has 12 letters.

Danish philosopher Søren Aabye Kierkegaard often wrote under different names so that he could argue with himself.

French philosopher Jean-Paul Sartre was so adamant that he didn't want the Nobel Prize for Literature that he hid in his sister-in-law's apartment until the prize-giving committee gave up trying to find him.

The first telephone directory to be published in book form was released in 1880 by the London Telephone Company. It listed 248 people but didn't give their numbers – to speak to one of them you had to contact the telephone exchange and ask.

In 1972, the academic journal *Applied Optics*, using the information from the Biblical chapter of Revelation that describes a lake in hell 'which burneth with fire and brimstone', calculated that the temperature of hell must be 444.85 degrees Celsius.

SOCIETY AND RELIGION

Their friends at *Physics Today*, using information in Isaiah regarding the lighting levels in heaven, went on to calculate that heaven's temperature was much cooler (as you might expect) at 231.35 degrees Celsius.

The Pope can't be an organ donor. Popes' bodies belong to the Vatican when they die, and they have to be intact.

CHAPTER 12.

POP CULTURE AND THE ARTS

(or Why do we like music?)

Why do we like music?

Charles Darwin described our ability to make music as a 'mysterious faculty'. But surely it's just a collection of sounds? How can it provoke emotion? Why, when we hear the opera *Carmen* or Status Quo's 'Rockin' All Over the World' being sung do we feel uplifted, excited even? And why, conversely, does the sound of a mournful violin playing the 'Theme from *Schindler's List*' create a sense of melancholy? It makes no evolutionary sense to be happy or sad when hearing music. Being able to detect B flat is not really a matter of life or death. Unsurprisingly, different theories abound. Some neuroscientists believe it might just be chance. The sounds of music coincidentally mimic other noises we have evolved to react to – such as a child crying or the screech of a predator, and this activates the same parts of our brain. Others think that sounds release the chemical dopamine into our brains. And dopamine makes us happy, which consequently means we choose to listen to music. It's a reward. Yet others think it sharpens our mental skills, such as emotion, and helps aid memory. Researchers studying sensory perception at George Mason University in Virginia, say: 'Humans also like regular patterns. We even look for these in stuff like clouds or puddles. When we find one we are pleased, and music offers many repeating patterns.' They add that: 'Discordant music is far less popular than melodic music, which backs up this theory.' The journal *Psychology Today* says: 'Music sometimes evokes nostalgia. We associate a tune with a good – or sad – part of our lives. And we may – or may not – want to repeat that.' If you stub your toe while listening to Brahms's Third Symphony, you might have damaged your relationship with the great composer forever. And finally, posit some, there's a shallower explanation. Apparently, the more attractive you find the performer the more attractive their music is to you.

Why do we sometimes end up with a tune playing repeatedly in our heads?

You know how it is. It's the middle of the night and you are desperate to sleep, but you can't get that tune you heard earlier out of your brain. It's called an earworm and neuroscientists think it has an evolutionary origin. Before humans could write, we sang songs to help people share and remember information. And melodies and rhythms provide cues for easier recall. The theory is backed up by the fact that usually we only replay 15 or 20 seconds of the piece in our brains, not an entire symphony or even a full pop song – we remember only the most catchy part, the bit that would be important to our illiterate ancestors. And it still works today. Some multi-storey car parks play different pieces of catchy music on each floor so drivers remember better which floor they are parked on.

Myth: We all like music.

Maybe not the same styles – your average fan of Mozart probably dislikes thrash metal (and vice versa) – but we all like some music, don't we? Well, people who have amusia don't. Their brains have a neurological deficit that leaves them unable to tell one piece of music from another. To them it's all just random sounds and they find it all quite irritating. We'll leave it to readers to make a rather weak joke here about the Spice Girls.

What are the most misheard song lyrics?

These, according to the *Hollywood Reporter* and *NME*:

'Purple Haze' by Jimi Hendrix:
'Excuse me while I kiss this guy' – actually 'Excuse me while I kiss the sky'.

POP CULTURE AND THE ARTS

'The Sidewinder Sleeps Tonite' by REM:
'Calling Jamaica' – actually 'Call me when you try to wake her'.

'Royals' by Lorde:
'You can call me green bean' – actually 'You can call me queen bee'.

'Country Boy' by Glen Campbell:
'Country boy, you've got your feet in a lake' – actually 'Country boy, you've got your feet in L.A.'.

'Livin' on a Prayer' by Bon Jovi:
'It doesn't make a difference if we're naked or not' – actually 'It doesn't make a difference if we make it or not'.

'You're the One That I Want' by John Travolta and Olivia Newton-John:
'I've got heels, they're made of plywood' – actually 'I've got chills, they're multiplying'.

'So Lonely' by the Police:
'Sue Lawley' – actually 'So Lonely'. (For the benefit of younger readers, Sue Lawley was a British newsreader and TV presenter at the time the song was released in 1978.)

'Jeremy' by Pearl Jam:
'Jeremy's smoking grass today' – actually 'Jeremy spoke in class today'.

'Africa' by Toto:
'There's nothing that a hundred men or more could ever do' – actually 'There's nothing that a hundred men on Mars could ever do'.

'Dancing Queen' by Abba:
'Feel the beat from the tangerine' – actually 'Feel the beat from the tambourine'.

'Blank Space' by Taylor Swift:
'All the lonely Starbucks lovers' – actually 'Got a long list of ex-lovers'.

We do, however, refuse to believe that when Bob Dylan sings 'The answer, my friend, is blowing in the wind', that some people hear 'The ants are my friends...'.

Myth: The Beatles were more popular than Jesus.

Well, that's what John Lennon said in 1966 when the Beatles arrived in the United States. And what a kerfuffle he caused. Protests from Christian groups followed the band wherever they played. Records were burnt and some American radio stations refused to play Beatles songs. Death threats were issued. But Lennon was wrong. If popularity is measured by how many people in the world had heard of the Beatles in 1966 compared with how many had heard of Jesus Christ, then the latter wins out by a ratio of about 5 to 1. However, some Beatles fans have argued that in 1966 the band sold 18 million albums of their music while only 15 million Bibles were bought. But, as some Christians also pointed out, many more people already owned Bibles so the comparison doesn't hold true.

Who is the richest music artist?

It depends how you calculate it. But Taylor Swift is the highest-grossing female live touring act of all time, the most streamed female act on Spotify and the first ever billionaire, of either sex or none, who has music as their main source of income.

Taylor Swift is beaten to highest-grossing live act by the Rolling Stones, U2 and Elton John (the highest-grossing solo music artist). It is calculated that the Stones have grossed US$2.2 billion (£1.75 billion) over six decades, while Swift has grossed US$2 billion in less than a third of that time, so she is catching up pretty quickly...

Of other solo female live acts, only Beyoncé and Madonna are in the top 20.

Myth: Bob Holness plays the saxophone on Gerry Rafferty's 'Baker Street'.

This is an enduring one. For years it was rumoured that Bob Holness, the besuited and sensible presenter of children's TV quiz show *Blockbusters*, played the haunting saxophone riff that defines Gerry Rafferty's hit 'Baker Street'. Writer and radio DJ Stuart Maconie apparently claims responsibility for this story gaining legs. When he worked at music newspaper *NME*, he made the claim in the publication's 'Believe it or Not' column, just to fill out space. But Maconie has a rival for the claim. DJ Tommy Boyd says that on his LBC radio show he ran a 'True or False' quiz, one of the questions being 'Did Bob Holness play sax on "Baker Street"?' The real saxophonist was Raphael Ravenscroft, who also lays claim to the myth. He says he was always being asked if he played the riff and got so bored with the question that he told people, no, it was Bob Holness.

Was *Barbie* the pinkest film ever?

Almost certainly. The pink even has its own Pantone shade: PMS 219C. The 2023 film led to a world shortage of pink. The film's production designer Sarah Greenwood used so much of paint company Rosco's ideal Barbie shade in her sets that it ran out.

The doll on which the film was based was introduced by toymaker Mattel in 1959. From the beginning it courted controversy. It was the first toy to be marketed directly to children via TV advertising. It was immediately criticised for its implausible and unrealistic figure. Researchers in Finland announced that if Barbie were a real woman she would be incapable of menstruating. As a consequence Barbie's body shape has been adjusted many times and in 2016 petite, tall and curvy Barbies were released, in addition to the original. Numerous 'career' Barbies have also been added to show she wasn't reliant on men to provide her income, including doctor, presidential candidate, pilot and astronaut.

Barbie's partner Ken was introduced in 1961. The children of Mattel's married co-founders were Barbara and Kenneth. The first African-American and Latino Barbies did not appear until 1980.

Although Barbie is positioned as the ultimate American woman, the doll has never been manufactured in the United States.

Myth: The first actor to play James Bond in a film was Sean Connery.

No. When the first Bond film, *Dr No*, was released in 1962 it opened with the now-trademark gun-barrel sequence where Bond is shown through the barrel of a gun before turning to shoot the would-be assassin as blood drips down the screen. The first actor to play this part was not Connery but a stuntman called Bob Simmons. Because this sequence was played before Connery appeared on screen a few moments later, Simmons is the first person to play 007 in a film.

Why does James Bond never die?

Estimates suggest that over all the 25 official Bond films, 007 has been shot at more than 10,000 times, making his chances of survival almost zero. However, the protective effect of hundreds of millions of dollars in box-office takings has created an impenetrable screen around him, meaning that he always scrapes through with just a few minor grazes.

Myth: But James Bond *is* dead.

Anybody who saw the latest Bond film, *No Time to Die*, knows that Bond was killed in the climactic scene. Well, if you believe that, you'll believe pretty much anything. As the last caption in the film's credits says: 'James Bond Will Return'.

Which film won the most Academy Awards?

Better known as the Oscars, the Academy of Motion Picture Arts and Sciences annual ceremony celebrates the best films and actors. Three films have won eleven awards at a single ceremony: *Ben-Hur* (1959), *Titanic* (1997) and *The Lord of the Rings: The Return of the King* (2003). Walt Disney, the animator and film producer, has won the most awards for a single individual (22 out of 59 total nominations). The most awards won by a single actor is four, by Katharine Hepburn, all for best actress.

POP CULTURE AND THE ARTS

Why do people get more money for pretending to be a nurse on a TV show than for being an actual nurse?

It's all to do with the free market. Films and TV shows cause more money to change hands (sales of the shows, endorsements, sponsorships, marketing, contracts). Conversely, while nursing is a skilled profession, it generates no revenue. Seems unfair because it is unfair.

Who said 'Play it again, Sam'?

Nobody. Well actually, lots of people since *Casablanca* was made in 1942, because lots of people misquote the film. It's attributed to actor Humphrey Bogart, but what he really says is: 'If she can stand it, I can. Play it!' while his co-star Ingrid Bergman says: 'Play it, Sam. Play "As Time Goes By".'

Other misattributed quotes:

'I disapprove of what you say, but I will defend to the death your right to say it.' Attributed to Voltaire but actually only a summary of his attitudes written by S. G. Tellentyre (Evelyn Beatrice Hall) in her biography of the philosopher.

'It is necessary only for the good man to do nothing for evil to triumph.' Attributed to statesman and philosopher Edmund Burke but not found in any of his works.

YAWNS FREEZE YOUR BRAIN

'The devil is in the detail.' Attributed to architect and academic Ludwig Mies van der Rohe. Ironic to relay that he actually said quite the opposite: 'God is in the details.'

'England and America are two countries divided by a common language.' Attributed variously to playwright George Bernard Shaw or prime minister Winston Churchill, but the nearest approximation is found in Oscar Wilde's *The Canterville Ghost*: 'We have really everything in common with America nowadays except, of course, language.'

'A lie can travel halfway around the world while the truth is still putting its boots on.' Attributed to writer Mark Twain, even though the attribution first appeared nine years after his death. Proof, it seems, that a lie can travel halfway around the world while...

'Warts and all.' Attributed to English Civil War parliamentarian and Lord Protector Oliver Cromwell (who was not of particularly picturesque appearance). What he really said to the man about to paint his portrait was: 'I desire you would use all your skill to paint my picture truly like me, and not flatter me at all, but remark all these roughnesses, pimples, warts and everything as you see me. Otherwise I will never pay a farthing for it.'

'The definition of insanity is doing the same thing over and over again and expecting different results.' Attributed to scientist Albert Einstein, who died in 1955, but the first recorded appearance of the quote is in literature produced for the organisation Narcotics Anonymous in 1981.

'Me Tarzan, you Jane.' Attributed to any number of actors who have played Tarzan of the Apes, but the words appear in neither the novel nor the films.

'You dirty rat.' Attributed to actor James Cagney, but not used by him in any of his films.

'Bigamy is having a wife too many, monogamy is having the same.' Attributed to Oscar Wilde, although the actual, rather sexist, source remains unknown.

QUICK-FIRE FACTS

The oldest musical instruments are nose flutes found in caves in Europe, dating back about 42,000 years.

The roar of the *Tyrannosaurus rex* in the film *Jurassic Park* is a blend of the sounds of an elephant, an alligator, a tiger and a penguin.

The terrifying shower scene at the start of the film *Psycho* took seven days to film. In the film it lasts 45 seconds.

In the film *Barbie*, actress Margot Robbie doesn't wear any rings, because the original Barbie doll on which the film is based had fused fingers that couldn't accommodate rings.

YAWNS FREEZE YOUR BRAIN

Sidney Poitier was the first Black man to win an Oscar for best actor, for his 1964 film *Lilies of the Field*.

All the clocks in the film *Pulp Fiction* are set to 4.20.

A horse fly that was discovered in Queensland in 1981 was, 30 years later, named after singer Beyoncé. Why? Because it has a striking golden tip to its abdomen, formed by a dense patch of golden hairs. Its taxonomic name is now *Scaptia beyonceae*.

Beyoncé has the most Grammy awards with 32.

Taylor Swift grew up on a Christmas tree farm in West Reading, Pennsylvania. (Explains why she has a song called, err, 'Christmas Tree Farm'.)

Former US president Barack Obama has more Grammys than Britney Spears. He has two, in the category of 'Best Spoken Word Album'. Britney only has one.

Ed Sheeran holds the record for the most concerts performed in a year. In 2009, he played 312 shows.

Justin Timberlake's mum was Ryan Gosling's legal guardian when Gosling was a child.

The Spice Girls – Scary, Baby, Sporty, Ginger and Posh – got their names from a journalist who couldn't remember their real ones.

POP CULTURE AND THE ARTS

In 2007, Rihanna insured her legs for US$1 million. Which is odd, considering she's a singer.

The first video ever aired on TV channel MTV was 'Video Killed the Radio Star' (how ironic) by the Buggles.

The animal rights organisation PETA once asked the Pet Shop Boys to change their name to Rescue Shelter Boys.

Morten Harket, lead singer of a-ha, holds the record for singing the longest single note in a pop song. In 'Summer Moved On', he sings a single note without stopping for 20.2 seconds.

Karaoke is Japanese for 'empty orchestra'.

Da da da daaaah. These are apparently the four most famous notes in classical music. The start of Ludwig van Beethoven's Fifth Symphony, in case you hadn't guessed.

Composer Johann Sebastian Bach had 20 children. Incredibly he still found time to write more than 1,000 musical works.

Wolfgang Amadeus Mozart composed 22 operas, 18 masses, at least 41 symphonies and 27 piano concertos – more than 600 works in total – and died aged 35. That's more than a symphony for each year of his life (plus all the other stuff).

YAWNS FREEZE YOUR BRAIN

By comparison, Beethoven wrote nine symphonies (he was 56 when he died) and Brahms wrote four (he was 63).

There are two skulls in composer Joseph Haydn's tomb. His head was stolen in 1809 and a replacement skull put in his tomb. In 1954, the real skull was returned but the substitute was never removed.

Composer Franz Liszt received so many requests for locks of his hair that he bought a dog and sent fur clippings instead.

The tension of the 230-plus strings in a grand piano exerts a combined force of 20 tonnes on its iron frame.

The world's oldest bookshop is Livraria Bertrand in Lisbon, Portugal. It was founded in 1732.

The first crossword was published in the *New York World* on 21 December 1913. It was compiled by British journalist Arthur Wynne. He called it a wordcross.

The longest currently running TV soap opera in the world is the UK's *Coronation Street*, broadcast continuously since 1960 (reaching well over 10,000 episodes).

However, *Coronation Street* is beaten to the overall soap endurance title by *The Archers*, a radio show that has been running daily on the BBC since 1951.

Paris Hilton is named after her parents' favourite city, and not after the Hilton hotel in Paris. However, she is the great granddaughter of Conrad Hilton, the hotel chain's founder.

CHAPTER 13.

CONFLICT AND OTHER NASTY STUFF

(or Why do humans fight wars?)

Why do humans fight wars?

Intuitively, you'd think that we'd do whatever we could to avoid such random death, destruction and misery. It's easy to blame governments for wars but it's notable how willing populations are to join in and offer support. Humans are, after all, just like any other higher-functioning animal, evolutionarily programmed to defend their families, territory and resources. The nineteenth-century American philosopher and psychologist William James suggested war is so commonplace because it has positive psychological effects. It creates a sense of unity and binds people together – not just the armed forces but the whole community – with a sense of cohesion and inspires citizens to behave honourably and unselfishly in the service of a

Paris Hilton is named after her parents' favourite city, and not after the Hilton hotel in Paris. However, she is the great granddaughter of Conrad Hilton, the hotel chain's founder.

CHAPTER 13.

CONFLICT AND OTHER NASTY STUFF

(or Why do humans fight wars?)

Why do humans fight wars?

Intuitively, you'd think that we'd do whatever we could to avoid such random death, destruction and misery. It's easy to blame governments for wars but it's notable how willing populations are to join in and offer support. Humans are, after all, just like any other higher-functioning animal, evolutionarily programmed to defend their families, territory and resources. The nineteenth-century American philosopher and psychologist William James suggested war is so commonplace because it has positive psychological effects. It creates a sense of unity and binds people together – not just the armed forces but the whole community – with a sense of cohesion and inspires citizens to behave honourably and unselfishly in the service of a greater good. Crucially, he believed, it surpassed the monotony of everyday life. So was James suggesting humans enjoy wars? No, in effect he was saying they relieve boredom. He discovered that the citizens of nations where the population was fulfilled in other ways – education, opportunities, careers and many forms of recreation such as arts and sports – were less likely to want to fight wars. But if such distractions are absent and there is perceived injustice (whether real or imagined), then war is a possible outcome. Sometimes, of course, there is a just cause, such as overcoming Nazi ideology in the Second World War. A nation will go to war if the benefits are deemed to outweigh the disadvantages, and there is no other mutually agreeable solution. More specifically, some have argued that wars are fought primarily for economic, religious and political reasons, plus territorial or ideological purposes, and sometimes for revenge. But again, in all these cases, governments need the support of their citizens, otherwise war would not happen. Perhaps disturbingly, research by Washington State University has shown that the prospect of inflicting retaliatory punishment triggers pleasure centres in the brain – the desire for revenge has led to some of human history's most infamous wars.

Were there any wars where nobody died?

Surprisingly, yes. There have been 14 wars where it is believed no combatants have died. We should perhaps arrange more of them. Perhaps the most unusual was the

Huéscar-Danish War, which began as part of the Napoleonic Peninsular War (1807–14). The Spanish Municipality of Huéscar declared war on the nation of Denmark in 1809 because it didn't agree with the Spanish government's decision to send 13,000 troops there to defend it. The 'war' lasted for 172 years because both sides forgot about it until it was rediscovered by a local historian in 1981, whereupon the mayor of Huéscar shook hands with Denmark's prime minister to bring the conflict to an end. There was an even longer bloodless war known as the 335 Year War, which began in 1651 and concluded in 1986. It was fought between the Netherlands and the Isles of Scilly, which can be found off the south-west coast of England. The Scillies had supported the Royalist forces in the English Civil War, while the Dutch had backed the Parliamentarians. When the Royalist navy retreated to the Scillies, the Dutch pursued them, declaring war, but when the Royalists surrendered shortly afterwards, the Dutch called off their attack. However, no peace treaty was ever signed until the Dutch ambassador to the UK visited the islands in 1986 to extend the hand of friendship.

Myth: The Scilly Isles and the Netherlands fought a 335-year war.

Okay, so we just said that they did, but maybe they didn't. Scilly Isles historian Rex Lyon Bowley wrote in 2001 that the Dutch Lieutenant-Admiral Maarten Tromp had no authority to declare war, while myth-busting historian Graeme Donald said neither side was sovereign: the

Scillies are part of the United Kingdom and Tromp wasn't a nation, so a state of war could not exist between them. It's still true, though, that nobody died. Which is nice.

The most recent bloodless war to conclude was the one between Canada and Denmark, who both laid claim to tiny Hans Island near Greenland. In 1984, Canadian soldiers visited the island, planted a Canadian flag and left behind a bottle of Canadian whisky. The Danish Minister of Greenland turned up later the same year with the Danish flag and a bottle of schnapps. Thereafter both nations took it in turns to plant their flags on the island and exchange bottles of booze. It became known as the Whisky War and was resolved in 2022, when a land border was drawn down the middle of the island and the two countries agreed to share it. So much more civilised than industrial slaughter.

How many nuclear weapons are there in the world?

The current estimate is 12,512, down from the highest recorded total of 70,300 in 1986. The following countries have or are believed to have, nuclear weapons: the United States, Russia, the United Kingdom, France, China, India, Pakistan, North Korea and Israel.

Myth: If all the nuclear weapons were detonated simultaneously, they would destroy the world many times over.

No, they probably wouldn't, but we can all agree that the outcome wouldn't be pretty. The planet would still exist as a spherical blob of rock but arguments reign as to how much life would continue to exist. The most positive estimates are that the death toll might only be about 80% of the global population, most of whom would have died from the initial explosions or from radiation poisoning or starvation within the first year, due to the collapse of global agriculture and its contamination from nuclear fallout. The nuclear explosion would also have unleashed a pulse of electromagnetic energy that would wreck everything from national power grids to anything powered by electronics or circuitry. Meanwhile, a huge volume of debris would be injected into the atmosphere, which would reduce the amount of sunlight reaching the Earth, producing a so-called nuclear winter with huge environmental impact. Nonetheless many species are resilient, either through genetics (rats, cockroaches) or through resourcefulness (humans) and, although life would be fairly miserable for quite some time, it would probably continue. The remaining human population would have to restart farming, which would require the removal of about 30 centimetres of contaminated top soil. More would die while this was happening or in regions where most of the weapons were used, but still a proportion of humanity would survive. So while it's possible life could be extinguished, it remains unlikely.

Has the nuclear deterrent worked?

The theory of Mutually Assured Destruction – the fact that if the US, Russia, China or any other nuclear power fired its missiles at another, they would respond in kind, thus destroying the planet (or not, see above) – states that no nuclear power will ever go to war with another. And this way the peace of the world is assured. But is that correct? The two arguments boil down to this:

Yes: No nuclear missiles have been fired in anger since 1945 and the world's military superpowers: the US, Russia (formerly the Soviet Union) and China have never been at war.

No: Since the end of the Second World War, there has not been a single day when war wasn't being fought somewhere on Earth.

What was the most deadly terrorist attack?

The Al-Qaeda attacks on the United States on 11 September 2001 were the deadliest in history – 2996 people were killed when terrorists flew planes into the twin towers of New York's World Trade Center and the Pentagon in Washington DC, while another attack was thwarted by passengers of a plane that crashed in Pennsylvania killing everyone on board.

What was the longest hijacking in history?

It depends whether we are talking time or distance. The hijacking of Kuwait Airlines Flight 422 on 5 April 1988 lasted 16 days, taking the plane from Bangkok to Algiers via Iran and Cyprus. However, the longest distance flown by a plane during a hijack was the 11,100 kilometres flown by TWA Flight 85, which was hijacked when it left Los Angeles bound for San Francisco on 31 October, 1969. It landed in Denver, New York and Maine before crossing the Atlantic to Shannon in Ireland and ending up in Rome.

The youngest hijacker in American history was 14-year-old David Booth, who hijacked a Delta Airlines flight from Cincinnati to Chicago in November 1969. Nobody died and Booth wasn't prosecuted because of his age. But even more brazen than Booth was a passenger who called himself D. B. Cooper. In 1971, Cooper hijacked a Northwest Orient Airlines flight between Portland, Oregon and Seattle. He told the crew he had a bomb and demanded $200,000 and a parachute. When the plane landed in Seattle he was given what he had requested. When the flight took off again he jumped from the plane over the Cascade Mountains and was never seen again (although some of the money was discovered in a creek nine years later).

CONFLICT AND OTHER NASTY STUFF

Myth: The sinking of the *Titanic* is the deadliest maritime disaster.

It might be the most famous, but after the RMS *Titanic* struck an iceberg and sank in the Atlantic Ocean in 1912, 'only' 1,513 lives were lost. When the German ship *Wilhelm Gustloff* was sunk by a Soviet submarine in January 1945, it is estimated that 9,343 people died.

Myth: The people of Hartlepool in north-east England killed a monkey during the Napoleonic Wars because they believed that's what French people looked like.

The Napoleonic Wars took place between 1803 and 1815, but the first mention of the killing was 40 years later, in a music-hall song of 1855, written by touring performer Ned Corvan, who wrote jokey songs about towns he was visiting. Corvan's story goes that the monkey was a mascot of French sailors in a ship that sank near Hartlepool and it washed up alive on the local beach. The people of Hartlepool had never seen a monkey before, nor, for that matter, a Frenchman. Mistaking its chattering for the French language of the enemy, they convicted it of being a spy and hanged it on the beach. But no French ships sank near Hartlepool during the wars and none of the ones that did, it seems, had monkeys aboard. And while at first the citizens of Hartlepool resented the implication that they were so naive, eventually they began to embrace the myth and it became part of local folklore. However, local historian Keith Gregson insists: 'There is absolutely zero evidence that it ever happened.'

Myth: The people of Hartlepool on the north-east coast of England elected a monkey as mayor.

Actually this isn't a myth, although it sounds as if it should be. Hartlepool Football Club's mascot is a monkey called, ummm, H'angus, and it was elected as mayor of Hartlepool, not once but three times. Underneath the monkey suit was Stuart Drummond, who ran for mayor, hoping to receive a bit of publicity for the struggling football team he supported. He never expected to win but the local population took to him and he served 11 years from 2002. Sadly, Mr Drummond only wore the monkey suit for campaigning and sometimes to please news photographers. And, of course, during Hartlepool's football matches.

QUICK-FIRE FACTS

The last land invasion of mainland Britain was when 1,400 soldiers of revolutionary France landed at Fishguard on 22 February 1797. They surrendered two days later.

The last Japanese soldier from the Second World War surrendered in 1974. He was living on the remote island of Lubang in the Philippines and hadn't realised the war had ended 29 years earlier.

CONFLICT AND OTHER NASTY STUFF

It might be apocryphal but it's said that Russia ran out of vodka celebrating the end of the Second World War, only 22 hours after victory was declared.

What most of the world calls the Second World War, Russians call the Great Patriotic War. So that's the war after which the vodka ran out.

The issue is confused by Russia also calling other conflicts Patriotic Wars, such as Napoleon's invasion of Russia in 1812 and what the rest of the world calls the First World War. It's not recorded how much vodka was drunk at their conclusions.

The First World War was known as the Great War (and still is) in many places until the Second World War came along.

The First World War was also known as the War to End All Wars, except, unfortunately, it didn't.

What Commonwealth countries call the American War of Independence is known in the United States as the American Revolutionary War.

When Japan launched its 73,000-tonne battleship *Musashi* in the port of Nagasaki, its displacement was so huge that it created a tidal wave more than a metre high, which swamped the city.

YAWNS FREEZE YOUR BRAIN

The youngest soldier to enlist for the British army in the First World War was 12-year-old Sidney Lewis. Somehow recruitment officers missed the fact that he was six years under the legal requirement.

During the Second World War, when London was being bombed, a dog called Juliana was awarded the Blue Cross Medal for animal acts of bravery, after she put out an incendiary bomb by urinating on it.

It is estimated that 4% of the sand on the beaches of Normandy is made up of shrapnel left from the D-Day landings of 1944.

In the Second World War possibly as many as 85 million people (the equivalent of the entire population of Germany today) were killed. Most were civilians in China and the Soviet Union.

The deadliest battle ever was the Battle of Stalingrad in 1942. 1.97 million people died.

The only casualties on United States soil during the Second World War were in Oregon, when five people were killed by a balloon bomb launched by Japan.

Tsutomu Yamaguchi survived both the nuclear bombings of Hiroshima and Nagasaki at the end of the Second World War, the only person to do so. He was working in

CONFLICT AND OTHER NASTY STUFF

Hiroshima on 6 August 1945 and then boarded a train home to Nagasaki on 8 August.

It is estimated that more than 37 million combatants, excluding civilians, have died in wars since 1800.

The last men to be hanged in Britain were murderers Gwynne Evans and Peter Allen on 13 August 1964. The death penalty in the UK was abolished on 8 November 1965.

The last woman to be hanged in the UK was Ruth Ellis, executed for the murder of David Blakely on 13 July 1955.

The last person to be hanged in the United States was Billy Bailey on 26 January 1996. Some states in the US still enforce the death penalty.

Since 1976, 143 prisoners on death row in the United States have been completely exonerated of the crime for which they were sentenced.

Murder rates rise 2.7% in the summer months.

The country with the highest homicide rate is Jamaica. In 2022, 53.34 people per 100,000 were murdered.

The country with the lowest homicide rate is Singapore, with 0.12 murders per 100,000 people in 2022.

However, Mexico beats Jamaica when it comes to total number of murders. In 2022, 33,287 people were murdered in Mexico.

Meanwhile, in Macau in 2022, only two people were murdered.

CHAPTER 14.

CLEVER SCIENCE, SILLY SCIENCE

(or Is necessity the mother of invention?)

Is necessity the mother of invention?

Only sometimes. If, for example, we are thinking of fixing global warming, then the many ideas put forward for capturing carbon show that the necessity of the problem of climate change has mothered any number of potential solutions. But, if we are being truthful, a vast number of things we take for granted today came about by accident. Israel's Arison Business School has calculated that of 200 inventions it deemed to be important to humanity, half of them were discovered before anybody had a clue what to do with them. Take Post-it Notes. In 1968 Spencer Silver of US conglomerate 3M was trying to come up with a super-powerful glue but failed, miserably. He found he'd invented a really weak one and parked the idea. However, a decade later, a colleague of his was looking for a bookmark and found a piece of paper with Silver's glue on it. He realised he could insert it into his book and it would stick in place without damaging the paper when he peeled it off. And now we all love Post-it Notes.

There's more glue stuff. In 1942 chemists at Eastman-Kodak in New York were trying to come up with a clear plastic for gunsights. They tried using cyanoacrylates, clear chemicals that thicken if water is added. But the chemicals simply stuck, very, very, firmly to anything they came into contact with. Useless! Until somebody figured out ten years later, after spilling some in a laboratory, that there was actually enough water in air to make them solidify. Despite his fingers being permanently stuck together, he'd invented superglue.

Who invented the periodic table?

Dmitri Mendeleev, a Russian chemist. It contains all the elements known to humans. The horizontal rows (or periods) are based on the atomic number of the element (the number of protons it contains), while the vertical columns contain elements of similar chemistry. When he drew up the table in 1869, there were 63 elements in it, now there are 118, but more will be discovered.

All the elements have a chemical symbol. For example, Ca is calcium. Most elements have a chemical symbol that is similar to their full name. However, the following elements don't: sodium (Na), potassium (K), iron (Fe), tungsten (W), silver (Ag), tin (Sn), antimony (Sb), gold (Au), mercury (Hg) and lead (Pb).

Ununtrium (Uut), Ununpentium (Uup), Ununseptium (Uus) and Ununoctium (Uuo) are the only elements to have a three-letter chemical symbol.

Why do we have a QWERTY keyboard?

One suggestion is that it slows the typist down, meaning that the old-fashioned typewriter keys on metal stalks didn't jam. Similarly, it's been argued that it keeps common letter pairs apart (but it doesn't because E and R – the second most common letter pairing in English – are right next door to each other). One last suggestion is that it enabled typewriter salesmen to show customers how simple the device was to operate by quickly typing 'TYPEWRITER QUOTE', just using the keys on the top row. However, it may just be that when typewriters were first being patented, Milwaukee resident Christopher Latham Sholes had devised a machine that looked more like a piano. He may simply have kept the same configuration of keys when he moved on to building the very first typewriter. Maybe we'll never KNWO.

Only a few English words can be written using the top row of a QWERTY keyboard, typewriter being one of them.

Who is Satoshi Nakamoto?

Satoshi Nakamoto is the pseudonym for the person (or maybe persons) who introduced the concept of Bitcoin in 2008 and 2009. Nakamoto remained active until about 2010 but has not been heard from since. If Nakamoto really does own the more than a billion Bitcoin they are rumoured to possess, they would be a multibillionaire today.

Two people identified as being Sitoshi Nakamoto have both denied it: Dorian Nakamoto, a physics graduate who worked on United States defence projects, and Nick Szabo, a computer engineer and legal scholar who created Bit Gold, a precursor to Bitcoin, in 2008. One other person, Australian scientist Craig Wright, has claimed to be Nakamoto but the High Court in London found overwhelming evidence against his claim. In July 2024 Wright finally admitted that he was not Nakamoto.

Is yellow fever infectious?

Yes it is, but Stubbins Ffirth, a doctor in nineteenth-century Philadelphia, believed it wasn't. And he set out to prove it by drinking the distinctive black vomit of patients with the disease. This was after he poured the vomit into incisions he'd made in his skin, and breathed in the vapours of heated vomit. Oh, and he smeared himself with the blood, saliva and urine of yellow fever patients too. It was after he'd survived these earlier experiments that he took the bold step of swallowing the undiluted puke. He survived and declared yellow fever to be transmitted by excessive noise and heat. He was wrong, yellow fever is contagious and deadly but, by sheer luck, Ffirth got away with it. One can only admire his dedication, all in the name of medical research.

One early cure for yellow fever was rubbing human fat on the skin. In Germany until the 1800s, leftover body parts were often used for medicine. Human fat was also sold as a remedy for broken bones, sprains and arthritis. Apothecaries regularly stocked human flesh, bones and fat, while skulls would be ground into powder to treat epilepsy.

Was Stubbins Ffirth the most reckless researcher?

Probably not. In 1898 at the Royal Surgical Clinic in Kiel in Germany, surgeon August Bier was trying to discover a strong anaesthetic that would work without rendering the patient unconscious. With selfless courage, his student August Hildebrandt volunteered to have his spine injected with cocaine. Bier then tickled Hildebrandt's feet – he felt nothing. But Hildebrandt was prepared to push medical science to its limits. He asked Bier to push a needle into his leg. Still nothing. So he requested a knife be stabbed into his thigh and a cigar stubbed out on his leg. Still Hildebrandt smiled. He then suggested Bier rip out his pubic and nipple hair (the nipples hurt which was gratifying because it showed the anaesthetic was staying local to his legs, as it was meant to do). After hammer blows to his shins and a hearty tug on his testicles, the two men deemed the test a resounding success. Whether Hildebrandt subsequently had any children remains unrecorded.

Barry Marshall was clearly the successor to Ffirth and Hildebrandt. And, in 2005, he won a Nobel Prize for his efforts. He believed that the *Helicobacter pylori* bacteria caused stomach ulcers in humans, which contradicted the view of the medical establishment, who believed only surgery could remedy ulcers. So Marshall swallowed some *Helicobacter pylori*, contracted an ulcer and cured himself with a simple course of antibiotics. His act overturned 100 years of medical opinion.

Why is it good to get things wrong?

There's the obvious cliché: we learn by our mistakes. But in science that's especially true. Stubbins Ffirth was only one in a long line of scientists – mad ones or otherwise – who got something wrong. But by doing so they all still advanced the cause of science. And that's because scientists experiment to rule things out just as much as they experiment to make a new discovery. They probably won't agree, but every patient who died on the operating table as a new medical procedure was performed, improved the chances of the next person surviving. We learn what works and what doesn't. Some scientists never discover anything new but their lifetime of dedication isn't wasted. For every researcher who discovers insulin successfully manages diabetes, there were a thousand others who discovered all the things that don't manage diabetes. And their contribution was just as valuable.

Surely though, not all science is good science?

It is fair to say that some things haven't really been thought through beforehand. During the Cold War the American military devised a shouting bomb. As it dropped from the sky on a parachute, it began to bellow recorded propaganda at those below, espousing US values of democracy, liberty and peace. And then, when it landed, presumably after converting all who heard it to its values of egality and freedom, it exploded, killing them all. After somebody pointed out this flaw it was never used.

So was that the most useless invention ever?

Again, probably not. South African designer Jan Louw devised an attachment for a vacuum cleaner that would cut your hair. You set it to your required length then ran the vacuum cleaner nozzle over your head, and hey presto! It didn't catch on because it hurt too much.

The Westinghouse Research Laboratory in Pittsburgh, Pennsylvania invented a coal-powered television. It too didn't catch on. Presumably the smoke it emitted obscured the screen.

And nor did the idea from Eugene Smyth of Winnipeg in Canada who devised a method of shrinking human corpses down to the size of dolls by removing all the liquid from them so that they would fit into cheaper, smaller burial plots. One imagines it also saved on pall-bearer expenses.

In 1993, officials at the United States Department of Energy suggested towing comets, which contain large amounts of ice, into Earth orbit, where their water could be extracted to help humans live in space without having to take their own supplies with them. It could also be used as propellant for spacecraft. Well, just because it hasn't happened yet doesn't mean it's a non-starter. Watch this, er, space...

What's the worst prediction ever?

Probably the one made by Albert Einstein's schoolteacher in Munich, who wrote in his school report that 'he will never amount to anything'. Einstein, of course, went on to formulate the theories of relativity and was, quite possibly, the cleverest person who ever lived (see page 184).

Other quite hopeless predictions:

'Television won't be able to hold on to any market it captures after the first six months. People will soon get tired of staring at a plywood box every night.' Hollywood producer Darryl Zanuck in 1946.

'The common cold will be only a memory by 2000.' US physician Lowry McDaniel in 1955.

'The horse is here to stay but the automobile is only a novelty, a fad.' An unnamed director of the Michigan Savings Bank in 1903, advising the Ford Motor Company's lawyer Horace Rackham not to invest his money in the business.

'The idea is idiotic.' William Orton, president of Western Union in 1876, discussing the future of telephones after their inventor Alexander Graham Bell offered him the patent.

'Electricity is just a fad.' Junius Morgan to his son J. P. Morgan of banking fame, in 1880. Fortunately, J. P. didn't listen.

'Heavier-than-air flying machines are impossible.' Lord Kelvin, mathematician, physicist and president of the British Royal Society in 1895.

'Before man reaches the Moon, your mail will be delivered within hours from New York to Australia by guided missiles.' Arthur Summerfield, United States Postmaster General under President Dwight D. Eisenhower in 1959.

'Should man ever succeed in going to the Moon, there would be little hope of his returning to Earth and telling us of his experiences.' *New Scientist* magazine, 1957. Doubtless some Moon-landing deniers (see page 34) still agree.

'There is no reason anyone would want a computer in their home.' Ken Olson, president, chairman and founder of Digital Equipment Corp in 1977.

'By the turn of the century, we will live in a paperless society.' Roger Smith, chairman of General Motors in 1986, one of about 20 million people to have made this prediction.

'I cannot conceive of any vital disaster happening to this vessel. Modern shipbuilding has gone beyond that.' Edward J. Smith, captain of the *Titanic* in 1912. He was backed up by Philip Franklin, vice-president of the ship's owner White Star Line, who said: 'There is no danger that *Titanic* will sink. The boat is unsinkable and nothing but inconvenience will be suffered by the passengers.'

'Everything that can be invented has been invented.' Charles H. Duell, Commissioner of the United States Office of Patents in 1899.

QUICK-FIRE FACTS

In 1997, a Buddhist monk in Japan set up a virtual temple to online files that had been deleted or lost.

Until 2002 the UK government reported on injuries suffered by the British population and their causes in its annual Home and Leisure Accident Surveillance System.

CLEVER SCIENCE, SILLY SCIENCE

In its final year it discovered that 13,132 Brits were injured by vegetables, 787 by sponges, 5,945 by trousers and 37 by tea cosies.

Football teams wearing red are more likely to win matches (but only if they are playing at home).

A metre is defined as the distance that light in a vacuum travels in 1/299,792,458 of a second. Previously it had been defined by a 1-metre platinum bar kept in Paris, but that could shrink or grow depending on the temperature.

Light takes about eight minutes to travel from the Sun to Earth.

A second is defined by the unperturbed ground-state hyperfine transition frequency of the caesium 133 atom. Obviously. (Or 1/86400 of a day, if that makes things easier).

In Albert Einstein's famous equation $E=mc^2$, E is energy, m is mass and c is the speed of light. If you want to know more, you'll have to take a degree in advanced physics.

The Egyptians were the first people to divide a day up into smaller chunks of time. Using sundials they divided daylight into 12 portions, meaning that summer 12ths were longer than winter 12ths.

In 2013, scientists in the US built an atomic clock so accurate that, had it been ticking since the Cambrian

YAWNS FREEZE YOUR BRAIN

geological period more than 540 million years ago, it would only be half a second out. No excuses for turning up late if you had one of those.

The first person to suggest that everything was made up of tiny pieces of matter called atoms – instead of combinations of earth, air, fire and water – was ancient Greek philosopher Democritus.

The first commercial radio broadcast made in 1920 will now have travelled around 100 light years from Earth. The Klingons probably know we're here.

42% of the rise in sea level is caused by water expanding because the planet is going through global warming.

Approximately 8 million tonnes of plastic enters the world's oceans each year (or the weight of more than 600,000 double-decker buses).

In March 2021, Croatian Budimir Šobat held his breath underwater for 24 minutes and 37 seconds. The average is usually around a minute.

More than 2 million tonnes of waste is sent to landfill sites every year.

A third of men and one-tenth of women in South Dakota have engaged in sexual activity while driving a car. Keep your eyes on the road, folks.

CLEVER SCIENCE, SILLY SCIENCE

Research shows that 1 in 2,500 pregnancies remain undiscovered until labour pains start.

Scientists at Brigham Young University carried out a study to discover what angle men need to urinate at to avoid their pee splashing back onto their shoes. It was 42 degrees downwards from the horizontal.

A 2018 study in the journal *Neuroradiology* discovered that wearing a tie can reduce blood flow to your brain by 7.5%.

If you are taking a photograph of 20 people, research by the Australian science organisation CSIRO has shown that you'll need to take it about six times before nobody has their eyes shut.

The fear of long words is called Hippopotomonstrosesquippedaliophobia. Its neologist obviously had a sense of humour.

The fear of palindromes is called Aibohphobia. Think about it...

Mirrors facing each other do not create infinite reflections. They absorb some of the light that hits them and each reflection gets darker than the previous one.

According to NASA, the science-fiction film with the least accurate science was *2012*, released in 2009. It depicts neutrinos from a solar flare heating up the Earth's core and ending all life while ignoring the fact the neutrinos pass straight through everything (including humans and the Earth) without harming anything.

A survey carried out at the turn of the millennium discovered that most people considered physicist Albert Einstein to be the cleverest person who ever lived. He was estimated to have an IQ of 160–190. Einstein himself, however, believed it to be Johann Goethe, the German author and polymath who founded the science of human chemistry and one of the earliest known theories of evolution. His estimated IQ is 225–250.

The person with the highest IQ is Adragon de Mello, an American born in 1976. His score is reportedly 400, although this has not been repeatedly and independently verified.

CHAPTER 15.

THE BEST OF THE REST

(or Why couldn't we fit this stuff in elsewhere?)

Why couldn't we fit this stuff in elsewhere?

This chapter was nearly called Bubbles, Liquids and Ice, and you'll find a fair few facts about all three hidden inside. But there was also lots of other quirky stuff that wouldn't quite fit into any of our other chapters, so here it is.

So what is a bubble?

Depends where it is. Below the water surface in the sea or in your bath it is a globule of gas surrounded by liquid. In the air it is called a membrane bubble and is a very thin film of liquid usually filled with air. When membrane bubbles are single they always form spheres to minimise their surface area – it's the shape that requires the least energy.

If you opened a fizzy drink in space and spilt some, the bubbles would remain inside it. On Earth they rely on gravity pulling the liquid downwards while they rise to the surface. But in space the liquid is not affected by gravity and so the bubbles go nowhere. However, you wouldn't actually have a fizzy drink in space. Astronauts are banned from drinking them because the body relies (again) on gravity to burp out the excess gas. And nobody likes a bloated astronaut.

Are bubbles always spheres?

No. When two membrane bubbles meet, they merge walls to, once again, minimise their surface area. If bubbles that are the same size meet, then the wall that separates them will be flat. If bubbles that are different sizes meet, then the smaller bubble will bulge into the larger bubble. This is because all bubbles have a higher internal pressure than the surrounding air and the smaller the bubble, the higher this pressure is. It's also why bubbles make a popping sound when they burst.

Why do membrane bubbles burst?

Three reasons, the most obvious being if they hit the ground or another object. But otherwise they eventually burst because either the liquid of which they are formed is pulled by gravity towards the bottom of the bubble leaving none at the top, or because the liquid evaporates. Adding soap to water slows these processes down by making the water more viscous.

At temperatures below about −25 degrees Celsius bubbles will freeze in the air and shatter when they hit the ground.

The best ingredients for making long-lasting membrane bubbles are: 85.9% water, 10% glycerol, 4% washing-up liquid, 0.1% guar gum (a food-thickening product).

There is such a thing as an antibubble. It is a spherical shell of air enclosing a droplet of water and forms when a water droplet falls into another body of water but retains the thin film of air that was surrounding it. So it is liquid, surrounded by air, surrounded by liquid.

The star-nosed mole and the American water shrew can smell underwater by rapidly breathing through their nostrils and creating a bubble that contains odour.

Myth: Water is wet.

It is and it isn't. And other things are wetter than water. Water is a liquid, we can guarantee you that. And it's made up of hydrogen and oxygen atoms, forming the molecule H_2O (that's two hydrogen atoms to one oxygen atom). In liquids these atoms are packed looser than they are in solids (but not as loose as they are in gases). When we touch a wet surface, these water molecules adhere to our skin. It's the sensation of this liquid layer on our skin that we call 'wetness'. But water itself can't be wet. Wetness requires a liquid to be in contact with a surface – your skin, or a cloth, or a pavement or many other things. So water can wet things that it touches but it can't wet itself. Its molecules also have bonds which hold it together, so although it spreads on surfaces it doesn't do this nearly as well as other liquids, such as alcohol, which get things wetter than water does. Wetness is the ability of a liquid to adhere to surfaces of solids so, for water to be wet, it needs to stick to something. And until that happens, it isn't really 'wet'.

Water can be treated with 'wetting agents' to make it wetter. These reduce its surface tension to make it spread further when it touches a surface.

Most substances become more dense when they solidify. But water is weird. It is at its most dense at 4 degrees Celsius, while it is still a liquid, and not at 0 degrees, when it becomes a solid (what we call ice). And this explains why ice (solid water) floats on (liquid) water. Most solids sink through their liquids, water doesn't.

Why do ice cubes from the tap sometimes have sediment in them?

That's because tap water isn't pure water, it contains small amounts of minerals from the rocks it passed through before flowing into the reservoir. When the water in your ice cube freezes, this tiny portion of minerals remains dissolved in what is left of the liquid water until the last water finally freezes. There is nothing left for the minerals to be dissolved in and they become solid once more, appearing as sediment at the bottom of your ice cube (or your gin and tonic).

In the 1980s, Japanese company Nippon Kokan invented a type of ice designed to make a satisfying crackling sound when put into drinks. If the ice was dropped into a glass of whisky, it was guaranteed to produce a 70-decibel crackle.

What happens at absolute zero?

Absolute zero – the coldest temperature that can be attained – is -273.15 degrees Celsius. All sorts of strange things are expected to happen if this temperature is ever reached. The motion of atoms will, it is believed, come to a complete standstill. However, the nearest anybody has got to it in a laboratory is about a billionth of a degree above absolute zero by scientists at the Massachusetts Institute of Technology.

Myth: The temperature of outer space is absolute zero.

It isn't, but it's not far off. It's generally regarded as being -270.45 degrees Celsius. But in 1995, the Boomerang

Nebula was discovered, which has a temperature that is even lower – only 1 degree above absolute zero.

Myth: Every single snowflake is different.

Well, it's what we've all been told since childhood. And it's almost certainly true. Except it might not be. Around a septillion (that's a 1 followed by 24 zeroes) snowflakes fall every year. But, when they first form, most snowflakes are just simple six-sided prisms. It's what happens to them afterwards that changes them into the intricate six-fold patterns they eventually become as they fall from the sky and encounter different weather conditions, temperatures, pressures and humidities. So if conditions remained identical from the moment of a snowflake's formation, in theory it could retain its original simple shape even as it falls to the ground. If this did happen, then it is possible that two snowflakes could be identical. Now just get searching through those septillion flakes...

Why does a feather fall at the same speed as a hammer?

In the late sixteenth century, astronomer and physicist Galileo Galilei suggested that two objects of different weight would fall at the same speed if dropped in a vacuum. In 1971, Apollo 15 astronaut David Scott was able to prove Galileo was correct when he carried out the experiment in a vacuum on the Moon's surface. Scott dropped a feather and a hammer at the same time and they fell at the same speed, landing on the lunar surface at the same time. The reason this happens is that gravity gives everything the same rate of acceleration despite its mass. The objects are said to be in free fall, so if they are dropped from the same height they will hit the ground together.

The feather came from Baggin, the United States Air Force Academy's mascot falcon and it is still on the Moon's surface, alongside the hammer. Nobody knows where the hammer was bought.

Myth: A feather falls at the same speed as a hammer.

We only said 'in a vacuum'. When an atmosphere is present, its 'thickness', or friction, acts on the feather far more than it does on the hammer. So if you drop them on Earth, the hammer will reach the ground first. Keep your feet well back.

Myth: When you sit on a park bench, you are sitting on a park bench.

What? Well, yes, you are, but you're not really. It's all to do with the fact that the atoms that make up every solid, gas or liquid in our universe never touch each other. They have electromagnetic repulsion – the closer they get to each other the more they are repulsed by the electrical charges of their constituent parts, in much the same way that two magnets repel each other. So although you think you are sitting on the bench, the atoms of your bottom (or more likely your trousers) are actually floating ever so slightly above the atoms of the bench.

Why do we hold our head in our hands when something goes wrong?

Allan Pease, an expert on body language, says we do this to replicate a mother holding a baby's head giving comfort and reassurance in stressful circumstances. It's universal across all societies. Other experts say it's our way of shutting out the world when something bad or embarrassing has happened. Your challenge after reading this book is to not put your hands on your head next time your football team misses an open goal. It's really, really difficult not to.

How much does wearing a chain or earrings slow down an athlete?

Not that much, you'd probably say. But if you want to have the best opportunity, you have to take what sports scientists call the 'marginal gains'. And it seems it pays to leave the jewellery behind. The margin of victory in a 100-metre sprint can be as small as 1 thousandth of a second. If you weigh 100 kilograms (or 100,000 grams) and your chain weighs 50 grams that's 0.05% additional weight you are carrying. If your 100-metre race takes 10 seconds, 0.05% of 10 seconds is 5 thousandths of a second.

Which is easily enough to relegate you from gold medal hero to fourth-placed zero. Ditch the bling.

And your hair can be a factor too. Cut it short. Analysis shows that when American Greg LeMond beat Frenchman Laurent Fignon in the final stage of the Tour de France in 1989, it was Fignon's hair that denied him glory. The final stage was a time trial. LeMond rode an aerodynamic bike while wearing an aerodynamic helmet. Fignon rode a standard bike and let his ponytail trail in the wind. It's estimated that the ponytail added slightly more than eight seconds to Fignon's time, which was exactly LeMond's winning margin, the closest in the tour's history. Interestingly, today all riders have to wear helmets, a stipulation that Fignon probably wished had been in place in 1989.

QUICK-FIRE FACTS

Some icebergs are formed from freezing seawater, yet their ice appears to be fresh-water ice and contains very little salt. This is because fresh water freezes at higher temperatures (0 degrees Celsius) than salt water (-1.8 degrees Celsius), so the first ice to form contains no salt. And off it bobs into the ocean.

Around 10% of the world's land surface and approximately 7% of its oceans are covered by ice.

YAWNS FREEZE YOUR BRAIN

The world's oldest ice is at the bottom of the ice covering Antarctica. It is around 1 million years old.

Despite being the coldest continent (-57 degrees Celsius on average inland), Antarctica receives very little snowfall (about 6.5 centimetres a year).

Conversely, it snows a lot more in the Arctic, even though there is far less land. Snowfall can be as much as 25 centimetres per annum.

Ice reflects 90% of all sunlight, meaning that if it all disappears, global warming will accelerate.

There is so much ice on Greenland in the northern hemisphere and the Antarctic in the southern, that it has a gravitational effect on the oceans, dragging the seawater towards them. If either lost their ice, seawater would be pulled in the opposite direction with potentially cataclysmic effects.

The most unstable naturally occurring element is francium. It has a half-life (the time taken for half the atoms in a sample to decay) of only 22 minutes, compared with the next least stable element, astatine, whose half-life is 8.5 hours.

Scientists estimate that there is as little as 20 grams of francium existing at any one time in the Earth's crust.

Smells can travel through liquids, as long as they are composed of soluble molecules.

Perhaps this should have been in our myth-busters. Science writer John Emsley once declared thulium to be the most useless element, saying you couldn't use it for anything. He was wrong – it can be used in portable X-ray devices.

Biologist Charles Darwin invented the office chair (or at least he was the first person to put wheels on the bottom of a normal chair, so he could slide around his study faster).

A lightning bolt – around 30,000 degrees Celsius – is five times hotter than the surface of the Sun.

An experiment in 2012 at the Large Hadron Collider in Switzerland created a temperature of 5.5 trillion degrees Celsius, the hottest temperature recorded on Earth and 366,000 times hotter than the centre of the Sun.

In 1990 Willie Jones contracted heatstroke and had a temperature of 46.5 degrees Celsius, the highest ever recorded in a human. He spent 24 days in hospital but survived.

The region of space around any star where conditions mean liquid water could be found on a planet, and therefore that planet could contain life, is known as the 'Goldilocks Zone', because the conditions are 'just right'. Earth is, of course, in the solar system's Goldilocks Zone.

YAWNS FREEZE YOUR BRAIN

Bubbles don't fly, they float, because they are less dense than the surrounding air.

The largest free-floating membrane bubble ever created had a volume of 96.27 cubic metres. It was made by Gary Pearlman of Cleveland, Ohio in 2015.

Trypophobia is a fear of closely packed holes.

It is illegal to own only one guinea pig in Switzerland in case it gets lonely.

A chef's hat has 100 pleats.

A jiffy is an actual unit of time (a nanosecond).

You can't hum if you hold your nose (and you just tried it, didn't you?)

In the course of a lifetime, a human eats about seventy insects while sleeping (some are very small).

A blob of toothpaste is called a nurdle.

CHAPTER 16.

WHAT'S GOING ON?

(or Why are we here?)

Why are we here?

Well, perhaps that's one best left to the theologians of whatever faith you subscribe to (or don't). But we do have a few ideas of how we got here, irrespective of preferred deity (or none). And most likely it's down to primordial soup – the hypothetical set of conditions present on the Earth around 4 billion years ago. But what was so special about the composition of that soup? What could have turned its infertile chemicals into a planet of teeming life? In 1953, at the University of Chicago, Harold Urey and Stanley Miller set out to create the atmospheric conditions of the early Earth. They filled a glass bowl with the raw ingredients of life – water, hydrogen, ammonia and methane – heated it to mimic the effect of the Sun and passed electric sparks through it to simulate lightning. Two weeks later the experiment had produced amino acids, the basic building blocks of proteins necessary for life. The experiment showed, at least in theory, how undetermined chemical reactions in the soup could lead to the first living cell, a process called abiogenesis. However, from that point on there are many different routes that could lead to the first lifeforms and palaeobiologists have argued long and hard over the most likely. Some propose a hypothetical stage that suggests life began with a simple RNA molecule that could copy itself without help from other molecules, eventually leading to DNA, the hereditary genetic material in humans and almost all other organisms. Other more outlandish theories exist, such as panspermia, the extraterrestrial bombardment of ancient Earth with meteorites carrying living organisms. The subject remains one of the most contentious in the field of life sciences.

How long does it take a human to decompose?

It depends where it is. In a vacuum it would last forever (or at least decay very slowly). But, assuming a dead body is exposed to the atmosphere, its cells break down almost immediately. This is called autolysis. Enzymes are released that begin to consume the corpse. If left alone it can begin to putrefy within 24 hours, thanks to the action of bacteria or other organisms, and the skin may show obvious signs of turning green as bacteria attack the gall bladder. Bacteria inside the body multiply and release gases that cause it to bloat (sometimes it can double in size) and smell. Between three and five days after death the skin blisters and organs decompose, leaking fluids from the body's orifices. Then the corpse will turn from green to a darker colour, starting the process known by the delightful term 'black putrefaction'. At two weeks, nails and teeth will start to fall out, and by the time a month has passed, the corpse will be liquefying into a dark, organic sludge. Assuming the body is placed into a coffin in the ground soon after death and no other life (such as insects or worms) can enter the coffin, then the flesh will decay, fall away and it will usually turn into a skeleton in about five to ten years. The skeleton itself can survive for decades depending on the conditions, but it too will eventually break down into dust. Obese humans tend to decompose quickly at first but the process slows if maggots gain access to the corpse because they prefer muscle to fat. Meanwhile, if a person had antibiotics or chemotherapy just before they died, decomposition is slowed down because both treatments kill off bacteria.

Who is the oldest preserved human?

Bodies have been preserved throughout human history, for cultural reasons or because the individual was considered important or famous enough to be embalmed, or even because some families of the deceased wished to keep the body on view. Embalming usually involves draining the blood and fluids from the corpse and replacing them with chemicals, in order to slow down bacterial decomposition. Usually this is for cosmetic purposes, so the body can be viewed at funeral homes or wakes. But many civilisations, such as the ancient Egyptians, have attempted to preserve bodies in perpetuity, filling them with oils, resins and myrrh. The oldest Egyptian mummy was discovered at Saqqara in 2023 – a man called Hekashepes who had been sealed in a limestone sarcophagus. He is believed to have died 4,300 years ago, or around 2,300 BC. However, he is beaten to the record of the oldest preserved intact human by Ötzi the Iceman who was discovered in the Alpine mountains between Austria and Italy in 1991. He was found frozen solid in the ice at an altitude of 3,210 metres. Archaeologists believe he was about 45 years old and have dated his remains to between 3,350 and 3,105 BC. It appeared the poor chap had been killed – an arrowhead was embedded in his shoulder. You can visit Ötzi. He is on display at the South Tyrol Museum of Archaeology in Bolzano, Italy.

From the 11th to the 19th centuries in Japan some Buddhist monks practised self-mummification. They sealed themselves in a tomb, ate a strict diet and meditated until death. The process could take more than three years.

Substances used to preserve bodies in an emergency have been diverse. Alexander the Great, who died in Babylon, was returned to Macedonia in a vat of honey – sugar is a great preservative. As is alcohol. After his death at the Battle of Trafalgar, Admiral Lord Nelson was returned to Britain in a barrel of brandy. He remained in it for two months until his state funeral, at which point it was noted that his body was in excellent condition. But please don't try drinking a whole cask of brandy as a means of protecting your internal organs, it won't work...

How has Lenin's corpse lasted so long?

The body of Vladimir Ilyich Ulyanov, better known to the world as the political revolutionary Lenin, has been on public display in the wall of the Kremlin in Moscow since 1924. It is in incredibly good condition, mostly due to the fact that much of its original biological matter has been replaced over time with plastics and other artificial materials by the Centre for Scientific Research and Teaching Methods in Biochemical Technologies in Moscow, who are responsible for keeping it in prime condition. The centre has also helped preserve the bodies of other prominent communist politicians, including North Korean premiers Kim Il Sung and Kim Jong Il, and Vietnamese leader Ho Chi Minh. If you want to be preserved for immortality, it seems you'd better take up left-wing politics.

Myth: The citizens of Pompeii were fossilised.

No, unlike what's going to happen to your body in the scenario above, they weren't. What happened when Mount Vesuvius erupted in the year 79 was a deadly cloud of superheated gas and hot ash rained down on Pompeii. The town's citizens, many of them caught in the open, were buried in the ash. But they weren't fossilised because the ash was so hot they were pretty much incinerated. As the molten ash hardened, the bodies quickly rotted away, leaving a cast in the ash of where they once were. And when archaeologists discovered these cavities in the 1860s, they poured plaster into them to reveal the figures of terrified, crouching citizens. So what you see if you visit Pompeii today are not fossilised humans, but plaster casts of where they once lay.

What is existentialism?

It's a tough question, but here goes. Because it's impossible to prove or disprove the existence of a God or Gods, existentialism helps to fill the gap. Existentialism encompasses a wide range of perspectives and its first proponents were nineteenth-century philosophers Søren Kierkegaard and Friedrich Nietzsche, and novelist Fyodor Dostoevsky. They proposed the idea that humans are born without purpose into an incomprehensible world – but each person has the ability to create his or her own sense of meaning and place. Our individual purpose and meaning is not given to us by deities, governments, teachers or other authorities. Of course, the fact that it's all down to us and us alone, with no input from outside forces, means that we may suffer what is known as an existential crisis, when we comprehend the challenges facing us in an absurd world. Similarly, such a belief can also teach us virtues such as courage and authenticity. For now, we'll have to settle for that. It's the best we can do under the circumstances.

Ancient Greek philosopher Diogenes believed in living a simple life. At one point he gave up all his possessions and spent some time living in a barrel.

To challenge society's understanding of why we feel shame, Diogenes urinated on people who annoyed him and defecated in public theatres. Today we'd probably give him an antisocial behaviour order...

Maybe it's true we are all just living in a computer simulation?

Well, the odds suggest that we are. The argument goes something like this. At some point humans will develop the capacity to write a computer program which, to the digital humans existing within it, will seem like reality. Others will copy the program. Within a short space of time there will be millions of simulations up and running, populated by billions of digital people, who all believe they are living in the real universe. So the chances of any of us actually being in the one and only real universe rather than one of the millions of simulations are so tiny as to be incalculable. And just as you are coming to terms with this new existence, the computer programmer will spill tea on his keyboard and blink you out of existence.

So what exactly *is* going on? What's the meaning of life?

Who knows? This book just reports the facts and tries to explain what existentialism is. And that's not easy! So we decided to close this last chapter by leaving the visionary, intelligent and imaginative meanderings on the meaning of life to the clever people below (apart, of course, from the quick-fire facts at the end):

'I think, therefore I am' – philosopher René Descartes.

'Irrationally held truths may be more harmful than reasoned errors' – biologist T. H. Huxley.

'The real question is not whether machines think but whether men do' – psychologist B. F. Skinner.

'Nothing is more dangerous than an idea, when you have only one idea' – philosopher Émile-Auguste Chartier, also known as Alain.

'Minds, like parachutes, only function when they are open' – whisky distiller Thomas Dewar.

'Keep an open mind, but not so open that your brain falls out' – attributed to, among many, professor Walter Kotschnig, physicist Richard Feynman and cosmologist Carl Sagan.

'The mind is its own place, and in itself, can make a heaven of hell, a hell of heaven' – poet John Milton.

'Ignorance is not innocence but sin' – poet Robert Browning.

'Science without religion is lame, religion without science is blind' – physicist Albert Einstein.

'The great tragedy of science – the slaying of a beautiful hypothesis by an ugly fact' – biologist T. H. Huxley (again).

'"You", your joys and your sorrows, your memories and ambitions, your sense of personal identity and free will, are in fact no more than the behaviour of a vast assembly of nerve cells and their associated molecules' – molecular biologist Francis Crick.

'Science is what you know. Philosophy is what you don't know' – philosopher Bertrand Russell.

'The fundamental cause of the trouble is that in the modern world the stupid are cocksure while the intelligent are full of doubt' – philosopher Bertrand Russell (again).

'To be, or not to be: that is the question' – playwright William Shakespeare.

QUICK-FIRE FACTS

166,859 people die, on average, every day. (That's 1.93 deaths every second.)

Bodies decompose four times as quickly in water.

The Wadi-us-Salaam cemetery in Najaf in Iraq is the world's largest. It covers 6.01 square kilometres and contains more than 6 million bodies.

It is estimated that by 2070 there will be more profiles of dead people on Facebook than living ones (10,000 are "created" every day).

The United States National Academies of Science estimates that more than 7,000 people a year die because of doctors' bad handwriting.

Cotard delusion, named after neurologist Jules Cotard, is a syndrome where you believe you are already dead even though you are clearly alive.

Around 500 people worldwide are cryogenically preserved in liquid nitrogen in the hope that biologists will one day be able to bring them back to life.

James Hiram Bedford, a professor of psychology at the University of California, was the first person to be cryogenically frozen in 1967.

There are more than 200 frozen corpses on Mount Everest.

A cremated body produces somewhere between 1.5 and 4 kilograms of ash.

'Six feet under', a slang term for 'buried', originated in London during the Great Plague of 1665, when the lord mayor decreed that, to prevent infection, that's how deep a plague-ridden corpse should be buried.

WHAT'S GOING ON?

The first legal cremation did not take place in the UK until 1885, in Woking, Surrey. Until then the church had objected to the practice of burning the dead.

The crematorium at Woking, Britain's first, was built seven years earlier and was tested by cremating a dead horse. But dead people had to bide their time until the government approved human cremation.

Up until the 1950s, the Fore people of Papua New Guinea ate the bodies of their relatives to cleanse their spirits.

The Wari people of Brazil also ate their dead relatives in order to process their grief (it's called endocannibalism). The Brazilian government banned the practice in the 1960s.

The Zoroastrian Parsi Community in India offers its dead to vultures rather than cremating or burying them.

Similarly, 80% of Tibetan Buddhists choose to have a sky burial, in which their bodies are placed on mountaintops and consumed by birds of prey.

Most animals don't eat their own species. Notable exceptions are cane toads, some spider species and caecilians (which must be quite difficult because they are sightless amphibians).

The Roman emperor Valentinian I reportedly died from a stroke provoked by angrily yelling at foreign ambassadors.

Meanwhile sixteenth-century Italian author Pietro Aretino is said to have died from laughing too much at an obscene joke during a dinner in Venice.

American lawyer Clement Vallandigham, defending a client accused of murder, accidentally shot himself while demonstrating how the victim might also have done so. He won the case though. Wonder who got his fee?

More people die from flying champagne corks than from poisonous spider bites.

More people die taking selfies than from shark attacks.

Approximately 2,500 left-handed people die every year using products designed for right-handed people.

Coffins are tapered at the base, caskets are rectangular.

ACKNOWLEDGEMENTS

Sally Manders
Thomas O'Hare
Cristina Amante
Melanie Green
Laura Fletcher
James Nightingale
James Kingsland
Eleanor Harris
Barney Harris
Polly Halsey
Alice Grandison
Rebecca Weigler
Claudia Bullmore
Emmy Pip Knight

Bedford Square Publishers is an independent publisher of fiction and non-fiction, founded in 2022 in the historic streets of Bedford Square London and the sea mist shrouded green of Bedford Square Brighton.

Our goal is to discover irresistible stories and voices that illuminate our world.

We are passionate about connecting our authors to readers across the globe and our independence allows us to do this in original and nimble ways.

The team at Bedford Square Publishers has years of experience and we aim to use that knowledge and creative insight, alongside evolving technology, to reach the right readers for our books. From the ones who read a lot, to the ones who don't consider themselves readers, we aim to find those who will love our books and talk about them as much as we do.

We are hunting for vital new voices from all backgrounds – with books that take the reader to new places and transform perceptions of the world we live in.

Follow us on social media for the latest Bedford Square Publishers news.

@bedsqpublishers
facebook.com/bedfordsq.publishers/
@bedfordsq.publishers

https://bedfordsquarepublishers.co.uk/